Museum of Guy's Hospital

Descriptive Catalogue of the Pathological Specimens

Contained in the Museum of Guy's Hospital: Volume II.

Museum of Guy's Hospital

Descriptive Catalogue of the Pathological Specimens
Contained in the Museum of Guy's Hospital: Volume II.

ISBN/EAN: 9783337015350

Printed in Europe, USA, Canada, Australia, Japan

Cover: Foto ©berggeist007 / pixelio.de

More available books at **www.hansebooks.com**

DESCRIPTIVE CATALOGUE

OF THE

PATHOLOGICAL SPECIMENS

CONTAINED IN

THE MUSEUM

OF

GUY'S HOSPITAL.

THIRD EDITION.

VOLUME II.

MORBID CONDITIONS OF THE LIVER,
PANCREAS, SPLEEN, SUPRARENAL BODIES,
URINARY ORGANS, AND MALE GENITALIA.

BY

LAURISTON E. SHAW, M.D. (Lond.), F.R.C.P.,
ASSISTANT PHYSICIAN AND CURATOR OF THE MUSEUM,

AND

E. COOPER PERRY, M.A., M.D. (Cantab.), F.R.C.P.,
ASSISTANT PHYSICIAN AND DEMONSTRATOR OF MORBID ANATOMY.

LONDON :
J. & A. CHURCHILL,
11, NEW BURLINGTON STREET.
MDCCCXCIX.

ALERE FLAMMAM.

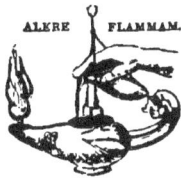

PRINTED BY TAYLOR AND FRANCIS,
RED LION COURT, FLEET STREET.

PREFACE.

In presenting this volume the writers desire to record their indebtedness to Mr. Targett for having placed at their disposal his descriptions of the Malignant Growths of the Bladder, which have with little change been incorporated in the text. Mr. Targett has also shared with the writers the responsibility of determining the nature of the histological changes in the great number of preparations described in this volume which have been submitted to microscopical examination.

It is believed that the abbreviated references to the Transactions of Societies and Periodical Journals will be easily understood by those using this Catalogue.

For the benefit of strangers to the Hospital it may be explained that *Insp.* refers to the Records of Post-mortem Inspections kept in the Curator's private room, and that *Surgical Reps.* and *Medical Reps.* refer to the Clinical Records kept in the Surgical and Medical Registrars' rooms respectively.

Guy's Hospital,
April 10th, 1899.

TABLE

OF THE

CONTENTS OF THE SECOND VOLUME.

SECTION XXV.—INJURIES AND DISEASES OF THE LIVER.

1246   **Hæmorrhage beneath the Capsule of the Liver.**

A marginal portion of liver mounted to shew its serous investment stripped from the organ and separated from it by a mass of blood-clot, in some places as much as an inch and a quarter thick. Below is mounted a portion of the liver, from which the coagulum has been removed to shew numerous superficial lacerations.

From Mary G., æt. 35, who, whilst in the ninth month of pregnancy, was seized after dinner on Christmas Day with violent vomiting, which persisted till her death thirty hours later. At the autopsy the whole of the convex surface of the liver was covered with a thick layer of coagulated blood, which appeared to have proceeded from a number of superficial lacerations in the organ. The rest of the viscera were normal, and there was no blood in the peritoneal cavity. *See Note Book*, No. 1, p. 156.

Presented by Mr. W. Smith, 1828.

1247   **Laceration of the Liver.**

A portion of liver mounted to shew upon its under surface a laceration extending a considerable distance into its substance. Over the laceration the capsule is seen to be intact.

Stephen O., æt. 41, was admitted under Mr. Cock, having been run over by an omnibus. He died a few hours after admission, and at the autopsy all the ribs were found to be fractured and the liver was almost completely divided by a vertical rent between the right and left lobes. *See Insp.* 1861, No. 161.

1248   **Laceration of the Liver.**

A portion of the right lobe of a liver, on the convex surface of which are seen several oblique lacerations involving the capsule and extending a short distance into the substance of the organ.

Elizabeth S., æt. 27, was admitted under Mr. Key in 1827, having fallen from a window. She died thirty-six hours after the accident, and at the autopsy the pelvis and several ribs on each side were found to be fractured. There was considerable effusion of blood into the peritoneal cavity and subperitoneal tissues. *See Insp.* vol. 4, p. 42.

1249   Laceration of the Liver.

A portion of the right lobe of a liver, shewing several lacerations, the edges of which are covered with recent lymph, a similar material being deposited upon the capsule of the organ.

James II., æt. 53, was admitted under Mr. Howse, having been crushed against a wall by a load which fell from a passing waggon. Bronchitis and pleurisy supervened, and he died six days after the accident. At the autopsy blood was found in the peritoneal cavity, and there was an encysted collection of bile-stained serum, measuring about two pints, between the liver and diaphragm. *See Insp.* 1890, No. 158.

1250   Laceration of the Liver in process of Repair.

The right lobe of a liver, shewing upon its convex surface an antero-posterior laceration six inches in length, and extending an inch and a half into the substance of the organ. There is a considerable quantity of recent lymph upon the capsule and between the edges of the rent. There is a second laceration upon the under surface of the liver.

Lyon B., æt. 29, was admitted under Mr. Hilton, having been injured by a roll of lead falling on to his chest. He suffered from hæmoptysis and dyspnœa, and died eight days after the accident. At the autopsy the 7th, 8th, and 11th ribs on the right side were found to be fractured, and the pleura was covered with recent lymph. There was a localized collection of bloody fluid measuring about one pint immediately beneath the liver. *See Insp.* 1859, No. 219 ; and *Drawing,* 348 (75).

1251   Laceration of the Liver in process of Repair.

The right lobe of a liver, on the convex surface of which are seen several irregular lacerations, some partially healed.

William B., æt. 32, was admitted under Mr. Poland, having been struck in the region of the liver by the handle of a machine.

Symptoms of pneumonia supervened, and the patient died eighteen days after the accident. At the autopsy the 8th and 9th ribs on the right side were found to be fractured, and there were numerous abscesses in both lungs. The peritoneum was blackened by effused blood. There was no general peritonitis. *See Insp.* 1864, No. 34.

1252   Liver Wounded in the Operation for Empyema.

The lower part of a right lung thickly coated with lymph and adherent to the diaphragm, together with the adjacent portion of the right lobe of the liver. Just below the lung, at its hinder part, is seen a ragged opening in the diaphragm leading to a deep laceration in the liver. There is recent blood-stained lymph on the surface of the liver around the margin of the laceration.

From James II., æt. 42, who suffered from an empyema following lobar pneumonia. An attempt was made to open the empyema by an incision in the ninth intercostal space. Three days later the patient died from purulent peritonitis.

1253   Needle in the Liver.

A portion of a liver, embedded in which is a needle an inch and a half long, its point projecting a short distance from beneath the capsule.

1254   Malformation of the Liver.

The liver of an infant, the quadrate lobe of which is separated from the right lobe, and occupies a lower position than normal. It is attached to the longitudinal fissure by a thin pedicle of fibrous tissue, and the gall-bladder is in contact with its posterior surface. The remaining lobes of the organ are normal.

Presented by Dr. Goodhart, 1889.

1255   Fissure in the Liver.

A portion of a liver shewing upon its upper surface a shallow linear depression three inches in length running obliquely backwards and inwards from the notch situated at the fundus of the gall-bladder. There are organised adhesions attaching the gall-bladder to the liver and duodenum.

Arthur B., æt. 48, was admitted under Dr. Pavy for gonorrhœal synovitis accompanied by persistent vomiting, from which three weeks later he died. He was not known to have received any abdominal injury. *See Insp.* 1880, No. 303.

1256   Deformed Liver.

A portion of a liver mounted to shew upon its upper surface a deep antero-posterior groove, corresponding to a ridge produced by the folding of the diaphragm upon itself. It is thought that the fold is due to the contraction of adhesions connecting the pleural surface of the diaphragm to the base of the lung.

John K., æt. 55, was admitted under Dr. Addison in 1828 for jaundice and delirium. He died seventeen days after admission, and at the autopsy the cæcal appendix was found to be ulcerated, and there were numerous abscesses in the liver. *See Insp.* vol. v. p. 140.

1257   Grooves upon the Liver.

A portion of the right lobe of a liver, shewing upon its upper surface four deep parallel grooves running in an antero-posterior direction.

1258   Liver Deformed by Tight-lacing.

A liver, presenting midway between its anterior and posterior margins a transverse depression running across the organ. At either end of this depression the lateral edges of the lobes are deeply indented. The antero-posterior diameter of the liver is considerably increased.

Bridget S., æt. 48, was admitted under Dr. Barlow for bronchitis, from periodical attacks of which she had suffered for twenty years. She died three weeks after admission, and at the autopsy the right side of the heart was found to be diluted and hypertrophied. *See Insp.* 1857, No. 28.

1259   A Deformed Liver.

A portion of the right lobe of a liver mounted to shew its anterior edge, for a distance of three inches to the right of the gall-bladder, turned upwards and backwards, so as to be in close contact with the upper surface. Along the line of flexure and for some distance above it the capsule covering the upper surface of the organ is somewhat thickened.

Elizabeth B., æt. 44, was admitted under Dr. Perry with intestinal obstruction, for which colotomy was performed. She died seven weeks after the operation, and at the autopsy a carcinomatous stricture was found in the sigmoid flexure, and there were distension ulcers in the cæcum and ascending colon. *See Insp.* 1891, No. 246.

1260   Compensatory Hypertrophy of the Spigelian Lobe.

A liver mounted to shew the lobulus Spigelii greatly enlarged and extending towards the left beneath and beyond the left lobe. It measures six inches transversely and five and a half inches from back to front. The greater part of the right lobe is occupied by several intercommunicating hydatid cysts.

William P., æt. 40, was admitted under Mr. Durham for an abdominal tumour with symptoms of intestinal obstruction. Laparotomy was performed and a hydatid cyst of the pelvis was opened and drained. The patient died the day after the operation, and at the autopsy general peritonitis was found. *See Insp.* 1878, No. 474.

1261   Atrophied Right Lobe of the Liver.

A portion of a liver, the right lobe of which is deformed and shrunken so as to measure about three and a half inches in its longest diameter. The left lobe is greatly enlarged, lardaceous, and slightly cirrhotic.

John O'N., æt. 40, was admitted under Dr. Rees for hæmoptysis, from which one week later he died. At the autopsy chronic phthisis was found in the lungs with tuberculous ulcers in the intestine. The kidneys were lardaceous. *See Insp.* 1857, No. 68.

1262 Atrophy of the Left Lobe of the Liver.

A portion of a liver, the left lobe of which is much shrunken, measuring two inches transversely and three inches from back to front. The shrunken lobe has a coarsely granular surface and histologically shews a considerable excess of fibrous tissue between the lobules. The right lobe, which is of normal size, presents several depressed scars upon its capsule.

Daniel M., æt. 55, was admitted under Dr. Pye-Smith for chronic phthisis, from which he died. At the autopsy the testes were found to be fibrous. *See Insp.* 1888, No. 389.

1263 Simple Atrophy of the Liver.

A portion of a liver, the capsule of which is slightly wrinkled. Histologically the structure of the organ appears to be normal.

Nahum B., æt. 53, was admitted under Dr. Barlow for ascites and œdema of the legs, having previously suffered from ague and dysentery in Hong Kong. He died a week after admission, and at the autopsy his body was much emaciated, and the liver weighed 32 ounces. *See Insp.* 1865, No. 324.

1264 Acute Atrophy of the Liver.

A very small liver, the capsule of which is loose and wrinkled. In the recent state the organ weighed nineteen ounces, and some parts of it were dark red and others of a bright yellow colour. Under the microscope the hepatic cells are seen to be replaced by an amorphous débris containing granules of bile pigment.

Ellen L., æt. 23, was admitted under Dr. Wilks for jaundice and vomiting, which persisted till her death. She had lately suffered from mental anxiety and was much depressed. She died eight days after admission, having been delirious during the proceding forty-eight hours. At the autopsy the liver was found lying contracted against the diaphragm, and the gall-bladder was empty. *See Insp.* 1859, No. 123.

1265   Acute Atrophy of the Liver.

A liver deformed by tight-lacing and considerably shrunken, the capsule being loose and wrinkled. In the recent state the organ weighed 25 ounces and was of a mottled red and yellow colour. Histologically there is disintegration of the hepatic cells and a catarrhal condition of the bile ducts.

Elizabeth B., æt. 30, was admitted under Dr. Barlow in a comatose condition and suffering from jaundice of four or five weeks' duration. Two days later she died, and at the autopsy the tubules of the kidneys were found to be filled with dark granular and biliary matter, and casts were detected in the urine in the bladder. *See Insp.* 1859, No. 125.

1266   Acute Atrophy of the Liver following Cirrhosis.

A portion of a liver, the under surface of which is wrinkled and nodular. On the reverse of the specimen the cut section shews the organ to be traversed by bands of white material. Histologically there is a slight excess of fibrous tissue between the lobules, a few of which appear to be normal, whilst in the greater number of them nothing can be seen but a structureless granular débris. In those lobules in which the change is less complete the disintegrating process is limited to the periphery.

George S., æt. 22, a man of intemperate habits, was admitted under Dr. Pye-Smith with jaundice of five weeks' duration, having suffered from hæmatemesis at the commencement of his illness. The hepatic dulness was almost entirely absent, and leucin was found in the urine. He passed into a comatose state, and died four days after admission. At the autopsy the liver was found to weigh 26 ounces, and its cut surface was of a brilliant yellow-ochre colour. *See Insp.* 1876, No. 233.

1267   Localised Red Atrophy of the Liver.

A section of a liver, the greater portion of which is smooth, firm, and dark in colour, whilst at the upper part of its convex surface is a considerable area, and elsewhere are small scattered foci, in which the hepatic tissue has a light yellow colour, and is of soft consistency. Histologically the structure of these parts appears to be normal, the darker parts shewing great dilatation of the capillaries, slight excess of fibrous tissue, and almost complete atrophy of the hepatic cells. Under the microscope a few miliary tubercles are seen scattered through the substance of the organ.

Wellington S., æt. 41, was admitted under Dr. Pye-Smith for phthisis associated with laryngitis and ascites. He died four weeks after admission, and at the autopsy the liver was found to weigh 57 ounces, and there were miliary tubercles in the suprarenal capsules. The kidneys were hard and congested. *See Insp.* 1894, No. 258.

1268   **Putrefactive Emphysema of the Liver.**

A portion of a liver, beneath the capsule and in the substance of which are numerous cavities varying in size from a pea to a millet seed. Histologically they are seen to be excavations in the hepatic tissue which is affected by post-mortem decomposition. *See Trans. Path. Soc.* vol. 35, p. 214.

Presented by Dr. Hale White, 1884.

1269   **Putrefactive Cysts of the Liver.**

A portion of a liver, the central parts of which present a sponge-like appearance, the result of putrefactive emphysema. Histologically the spaces are seen to be excavations in the normal liver substance with no definite lining membrane. *From the Fort Pitt Museum.*

1270   **Putrefactive Cysts of the Liver.**

A section of a cirrhotic liver, mounted to shew numerous minute vesicles scattered throughout the organ. Histologically the vesicles are seen to be cavities without definite lining membrane, and appear to have been produced by decomposition.

From the Fort Pitt Museum.

1271   Fatty Liver.

A portion of a liver, with a smooth surface and of soft
consistency. The section has a homogeneous appearance, and histological examination shews that nearly all
the hepatic cells are loaded with fat.

From Mr. W., who had suffered from syphilis and dysentery,
and died from a fæcal abscess which opened externally above the
crest of the ilium. At the autopsy the abscess was found to be
due to a stricture of the large intestine, and there was a second
stricture in the rectum. *See Insp.* vol. 2, p. 90.

1272   Fatty Liver in Phosphorus Poisoning.

A section from the right lobe of a liver, mounted to
shew the condition of extreme fatty degeneration. In
the recent state the organ was pale in colour, and so
light as almost to float in water.

Alice G., æt. 18, was admitted under Dr. Goodhart, having
swallowed phosphorus paste, and died seven days after the
accident. Whilst in the hospital she vomited frequently, had
severe abdominal pain, and hæmorrhage from the mouth and
gums. The liver was felt to be enlarged. At the autopsy
fatty degeneration was found in the muscles, and in the epithelium of the kidneys. *See Insp.* 1891, No. 35.

1273   Lardaceous Liver.

A portion of a liver, the substance of which is converted
into a firm, translucent, almost homogeneous material,
which, under the microscope, is seen to be stained rosy
red when treated with methyl violet.

James M., æt. 5, was admitted under Mr. Morgan in 1827 for
disease of the vertebræ, and died two years after admission. At
the autopsy the liver was found to weigh 54 ounces. *See Insp.*
vol. 2, p. 54.

1274   Lardaceous Liver.

A portion of a liver treated with iodine to shew the
lardaceous material in it stained of a dark mahogany
colour.

John R., æt. 10, was admitted under Mr. Key in 1832 with an abscess about the hip joint, and died two years later from hæmorrhage from the femoral artery. *See Insp.* vol. 19, p. 28; and *Preps.* 1317 (40) & 1504 (80) (2nd edit.).

1275   **Lardaceous Liver.**

A marginal portion of liver, the cut surface of which has a translucent appearance from the presence of lardaceous material. The yellow area depressed below the surface of the organ corresponds to a caseous deposit, parts of which have undergone calcareous degeneration.

James S., æt. 24, was admitted under Dr. Habershon for spinal caries associated with a sinus in the right groin which had discharged pus for twenty years. There was general dropsy with considerable ascites, for which paracentesis abdominis was performed. He died two days after admission, and at the autopsy the liver was found to weigh 163 ounces. The intestines and kidneys shewed lardaceous change, and there were a few tuberculous nodules in the epididymis. There was no evidence of syphilis. *See Insp.* 1875, No. 164; and *Trans. Path. Soc.* vol. 27, p. 324.

1276   **Lardaceous Liver from Syphilis.**

A portion of a liver, the surface of which presents several depressions, and is covered by a thickened pitted capsule. . The section shews slight excess of fibrous tissue and numerous areas where the hepatic cells have been converted into lardaceous material.

Thomas II., æt. 43, was admitted under Dr. Addison with ascites, œdema of the legs, and enlargement of the liver and spleen. He was of intemperate habits and had had syphilis. Nineteen days after admission he died, and at the autopsy a subperitoneal hæmatoma was found around a puncture made in paracentesis abdominis. The liver weighed 76 ounces and the spleen 20 ounces. The kidneys were lardaceous, and there were numerous ulcers in the colon. *See Insp.* 1857, No. 58; and *Prep.* 1193.

1277   **Lardaceous and Cirrhotic Liver.**

A portion of a liver, the cut surface of which shews towards the periphery of the organ numerous trans-

lucent areas about a line in diameter, separated from
each other by thin strands of fibrous tissue. Histologically these areas are seen to consist of hepatic
lobules which have undergone an extreme lardaceous
change. The surface of the organ is granular from
cirrhosis.

Peter C., æt. 55, was admitted under Dr. Barlow for jaundice
and emaciation, and died eight weeks after admission. At the
autopsy the liver was found to weigh 93 ounces, and lardaceous
change was observed in the spleen, kidneys, and alimentary canal.
See Insp. 1865, No. 315.

1278   Lardaceous and Fatty Liver.

A portion of a liver, the cut surface of which presents
a mottled appearance, the lighter parts, which are at
the periphery, representing areas of fatty infiltration,
whilst the darker are seen under the microscope to have
undergone considerable lardaceous change.

Henry G., æt. 25, was admitted under Mr. Cock for abscesses
and fistulæ resulting from hip-joint disease. About four months
later he died, and at the autopsy the liver was found to be enlarged, and there was much serous fluid in the peritoneal cavity.
The spleen was slightly lardaceous. *See Insp.* 1864, No. 168.

1279   Lardaceous Liver.   Perihepatitis.

A portion of a liver with the diaphragm, to which its
thickened capsule is firmly united. The cut surface of
the organ is studded with irregular translucent areas,
which, under the microscope, are seen to consist of
lardaceous deposit replacing the hepatic lobules.

Emily R., æt. 28, was admitted under Dr. Rees for general
anasarca and albuminuria. She had previously suffered from
syphilis, and twelve months before admission had been tapped for
ascites. Two months later she died, and at the autopsy the liver
was found to weigh 70 ounces and to be so deformed by the contraction of its thickened capsule as to present a cylindrical shape.
The intestines, suprarenal capsules, and kidneys were highly
lardaceous. The peritoneum was acutely inflamed. *See Insp.*
1869, No. 64.

A 1279   Lardaceous Liver with localised Calcification.

A portion of a lardaceous liver, at the hinder part of which is seen, immediately beneath the capsule, a mass of dense yellow material, in which are embedded several whitish calcareous nodules. Histologically the mass consists of hepatic tissue exhibiting the appearances of advanced cirrhosis undergoing partial calcification.

Thomas G., æt. 18, was admitted under Dr. Gull for anæmia associated with dropsy and albuminuria, and died three months later. At the autopsy sarcomatous growth was found in the mesenteric glands and in the kidneys, the latter as well as the spleen being affected by lardaceous disease. *See Insp.* 1864, No. 49; and *Prep.* 1236.

1280   Nutmegged Liver.

A portion of a liver, the cut surface of which has an appearance somewhat resembling a section of a nutmeg. Histologically the dark brown areas are seen to consist of dilated radicles of the hepatic vein in the centre of the lobule, the paler peripheral portions being composed of fatty and atrophied liver-cells.

Richard M., æt. 47, was admitted under Dr. Pavy with signs of cardiac hypertrophy and dilatation, from the effects of which he died. At the autopsy the heart was found to weigh 23 ounces, and the aortic valves were thickened and adherent. The spleen and kidneys were indurated. *See Insp.* 1879, No. 240.

1281   Nutmegged Liver.

A portion of the right lobe of a liver, partially injected, and shewing its serous surface for the most part to be thickly covered with minute tags of fibrous tissue. On the anterior border and adjacent under surface the capsule of the organ is, as it were, sanded over with granules which are considerably smaller than miliary tubercles. Histologically the hepatic tissue presents the appearances produced by prolonged congestion of the intralobular vein, and the projections from the capsule are seen to consist of organised fibrin.

William R., æt. 24, was admitted under Dr. Taylor for cardiac dropsy of several months' duration, for the relief of which his legs were punctured. Cellulitis supervened, and the patient died four days after admission. At the autopsy the heart was found to weigh 24 ounces; the pericardium was universally adherent, and the aortic and mitral valves were fibrous and incompetent. The liver weighed 61 ounces and was nutmegged. There were infarcts in the lung, and the viscera generally were congested. *See Insp.* 1894, No. 244.

1282   Nutmegged Liver with Slight Cirrhosis.

A portion of the anterior border of a liver, the surface of which is uneven and sprinkled over with very minute fibrous tufts, which histologically are found to contain blood-vessels. The branches of the hepatic vein are filled with thrombus, and under the microscope the capillaries are seen to be dilated, the liver-cells atrophied, and the connective tissue somewhat increased in amount.

James W., æt. 7, was admitted under Dr. Gull for enlargement of the liver and spleen. Three weeks after admission ascites supervened, and at the time of death, which occurred three weeks later, the abdomen was greatly distended, and there was enlargement of the superficial veins and slight œdema of the legs. At the autopsy the liver was found to be hard and its surface irregular. The branches of the hepatic vein at its termination in the cava were many of them narrowed, and one of them actually obliterated. Several contained a firm clot "which was not old nor altogether quite recent." The heart and kidneys were healthy. *See Insp.* 1862, No. 31.

1283   Perihepatitis.

A liver, the surface of which is irregularly nodulated and covered with a thick fibrous capsule, upon which is seen a granular deposit of recent lymph. The organ is much contracted, and its sharp anterior margin rounded off.

1284 Localised Capsulitis of the Liver.

A portion of a liver, mounted to shew an irregular fibrous plate beneath its capsule measuring three inches in length, half an inch in width, and extending a quarter of an inch into the substance of the organ. Histologically the plaque has the structure of fibrous tissue.

1285 Fibrous Plate in the Capsule of the Liver.

A portion of the liver, beneath the capsule of which is seen embedded a mass of cartilaginous material three inches long and one inch broad, extending half an inch into the substance of the organ. Its surface, which is smooth, is slightly raised above the surrounding capsule, and its margin is well-defined. Histologically the deposit is seen to consist of hyaline fibrous tissue.

William S., æt. 45, was admitted under Dr. Bright in 1832 and died from valvular disease of the heart. *See Insp.* vol. 12, p. 5.

1286 Fibro-calcareous Nodule embedded in the Liver.

A portion of a liver, presenting upon its surface a deep depression, in which was lodged the fibro-calcareous nodule mounted below.

1287 Fatty Nodule in the Liver.

A portion of a liver, partly embedded in the surface of which is seen a firm white oval nodule about a third of an inch in diameter. Below is mounted a similar nodule which was found attached to the liver and has been incised to shew its structure. Histologically it consists of a fibrous envelope enclosing a fatty nucleus

1288 Fatty Nodule in the Liver.

A portion of a liver, shewing upon its surface an oval nodule three-eighths of an inch in its longest diameter, partly embedded in the substance of the organ. A section has been made through the nodule, and histological examination shews it to consist of a fatty nucleus surrounded by a fibrous capsule about a sixteenth of an inch in thickness.

William R., æt. 71, was admitted under Mr. Hilton with an enlarged prostate and retention of urine, and died four days later from hypostatic pneumonia and interstitial nephritis. *See Insp.* 1855, No. 162.

1289   Abscess of the Liver.

The right lobe of a liver, with the kidney of the same side, shewing on the lateral aspect of the former a part of the cavity of a large abscess. Its walls are formed above by an excavation in the liver, and below by thickened peritoneum. In the recent state the abscess passed behind the peritoneum into the pelvis and presented in the perinæum.

Eliza C., æt. 31, was admitted under Dr. Babington in 1804 with pain and tenderness in the right side, emaciation, and fever. Her symptoms were attributed to catching cold during menstruation nine months previously. She occasionally passed blood by stool, and died eighteen days after admission. At the autopsy six pints of pus were found in the abdominal abscess. *See Old Museum Book*, No. 5.

1290   Abscess of the Liver.

A portion of the right lobe of a liver, incised to shew the cavity of a large abscess, the walls of which are exceedingly irregular and ragged.

From Henry W., æt. 39, a man of intemperate habits, who was admitted into an asylum for melancholia, and three weeks later was seized with severe pain in the right hypochondriac region. His liver was observed to be enlarged. He passed into a typhoid condition, and died eight days after the onset of acute symptoms. At the autopsy the abscess in the liver was found to contain one pint and a half of pus. The alimentary canal was free from ulceration. *See Note Book*, No. 2, p. 43.

Presented by Messrs. Carrington and Gale.

1291   Abscess of the Liver.

The right lobe of a liver, with the kidney, shewing in the former the ragged cavity of an abscess, a portion of the wall of which is formed by the anterior surface of the kidney. In the recent state the abscess was of the " size of a child's head."

Jane M., æt. 19, was admitted under Dr. Rees with pain in the right side and dulness at the bases of both lungs. She suffered from diarrhœa, and died four and a half months after the onset of her illness. At the autopsy the colon was found to be extensively ulcerated. *See Insp.* 1867, No. 335.

1292   Abscess of the Liver.

A portion of the right lobe of a liver, shewing towards its upper surface, to which the diaphragm is adherent, the cavity of an abscess communicating with a large branch of the hepatic vein, the main trunk of which is indicated by a blue rod. Below is mounted a slough which, together with five ounces of pus, formed the contents of the abscess.

William II., an adult, was admitted under Dr. Rees with rigors and symptoms of pulmonary disease. He died about a fortnight after his admission, and at the autopsy the lungs were found to contain numerous abscesses and there was a thrombus in the inferior vena cava. In the cæcum were several old and recent ulcers. *See Insp.* 1872, No. 205.

1293   Abscess in the Liver with Dysentery.

A portion of the right lobe of a liver, to the surface of which the diaphragm and lung are firmly adherent. The liver has been incised to shew at its upper part the cavity of an abscess measuring four and a half by three inches. The interior of the cavity presents a trabeculated appearance, and its walls are formed of a dense layer of fibrous tissue about half an inch in thickness. A blue rod indicates the opening through which the abscess was drained.

John D., æt. 23, who about six months previously had suffered

from dysentery in Calcutta, was admitted under Dr. Pavy with an
enlarged liver. Ten days later thirty ounces of pus were evacuated
by an incision in the ninth intercostal space below the angle of the
scapula. He died about six weeks after the onset of his illness,
having been hemiplegic for the two days preceding his death. At
the autopsy the colon was found to be severely ulcerated, and there
were abscesses in the brain. *See Insp.* 1883, No. 354.

1294   Abscess of the Liver following Scarlatina.

The right lobe of a liver, laid open to shew the ragged
interior of a large abscess, which in the recent state
contained about a pint of pus. There is recent lymph
upon the capsule of the organ.

Julius S., æt. 2, was admitted in 1830 with scarlet fever, and
died fifteen days after admission. At the autopsy the stomach,
intestines, and gall-bladder appeared healthy. *See Insp.* vol. 14,
p. 131.

1295   Hepatic and Perisplenic Abscesses.

The left lobe of a liver, with the spleen and portions of
the lung, diaphragm, and ribs, mounted to shew two
abscess-cavities, one situated wholly within the liver
and the other bounded by adhesions between the spleen
and neighbouring organs. The latter has destroyed
the substance of the liver, but has no direct communication with the abscess within the organ.

1296   Pyæmic Abscesses of the Liver.

A portion of a liver riddled with abscesses varying in
size from a line to half an inch in diameter.

James D., æt. 37, was admitted under Mr. Hilton with a
whitlow, for which the finger was amputated. Pyæmia supervened, and the patient died 16 days after the operation. At the
autopsy abscesses were found in the lungs, and the shoulder-joint
contained pus. *See Insp.* 1856, No. 120.

1297   Pyæmic Abscess of the Liver.

A small portion of liver, shewing beneath the capsule
an abscess of irregular shape measuring one inch and a

half in its longest diameter. Its interior is roughened and communicates with a branch of the portal vein, which in the recent state contained pus.

John W., æt. 40, was admitted under Mr. Hilton for cellulitis of the arm following an injury. He died of pyæmia four weeks after the accident, and at the autopsy multiple abscesses were found in the liver, lung, and kidneys. There was acute peritonitis, attributed to the rupture of one of the hepatic abscesses. See Insp. 1854, No. 115.

1298 Pyæmic Abscesses of the Liver.

A portion of a liver, shewing several small abscesses with crinkled walls, and in their neighbourhood a large branch of the hepatic vein containing a canalised thrombus, the proximal end of which is smoothly rounded off.

John M., æt. 29, was admitted under Mr. Hilton for a suppurating bubo, and died 18 days later from pyæmia. At the autopsy multiple abscesses were found in the lungs and liver. See Insp. 1868, No. 4.

1299 Multiple Abscesses in the Liver.

A portion of the right lobe of a liver, the cut surface of which shews numerous excavations varying in size from a pin's head to a pea and partly filled with inspissated pus. Histologically the capsule of Glisson is found to contain an excess of small round cells and in many places to present a minute abscess, in the centre of which is a mass of orange-coloured bile-pigment.

William B., æt. 39, was admitted under Dr. Goodhart with febrile jaundice of ten days' duration, having twice previously suffered from biliary colic. Vomiting and diarrhœa supervened, and the patient died ten days after admission. At the autopsy a calculus was found partially obstructing the common bile-duct, and the gall-bladder was distended, but neither ducts nor gall-bladder appeared to be ulcerated. See Insp. 1888, No. 457; and Guy's Hospital Gazette, 1888.

1300   Calcified Wall of Hepatic Abscess.

Numerous calcareous plates of irregular shape which were removed from the wall of an abscess-cavity in the liver.

Antonio M. was admitted under Dr. Bright in 1831 with a distended abdomen and dilatation of the superficial abdominal veins. At the autopsy an abscess was found in the right lobe of the liver pressing upon the inferior vena cava. *See Insp.* vol. 10, p. 160.

1301   Abscess of the Liver invading the Lung.

A portion of a liver and diaphragm, with the lower lobe of the right lung, firmly united to each other by inflammatory tissue. In the liver there is a large abscess with shaggy walls, which has perforated the diaphragm and excavated the base of the lung.

1302   Cirrhosis of the Liver.

A portion of a liver, the surface of which is broken by flattened nodules varying in size from a line to half an inch in diameter. The capsule of the organ is of normal thickness. The cut surface shews circular areas of various sizes surrounded by a dense white tissue. Histologically the hepatic lobules are surrounded and separated from each other by an excessive amount of fibrous tissue, and branches of the portal vein are filled with collections of large spheroidal cells resembling cancerous thrombi.

John B., æt. 52, was admitted under Dr. Pavy for ascites of one month's duration associated with œdema of the legs. Three weeks after admission he was tapped, and 15 pints of serous fluid were withdrawn. Three days later he died, and at the autopsy the liver, which weighed 36½ ounces, was found to be uniformly affected in the manner shewn in the preparation. The trunk of the portal vein and some of its branches in the substance of the liver were filled by adherent softening thrombus. No appearance of malignant disease was seen in any organ. *See Insp.* 1876, No. 409; and *Trans. Path. Soc.* 1877, p. 137.

1303   Cirrhosis of the Liver.

Portions of a liver, the greater part of which is converted into white fibroid tissue, in which are seen yellow islets composed of degenerated hepatic cells. The thinner section shews an oval area measuring about two inches in its longer diameter, surrounded by a fibrous capsule and occupied by hepatic tissue, the cells of which are seen under the microscope to be in a more advanced stage of degeneration.

From a Swede, æt. 54, who died under the care of Dr. Carrington in the Seamen's Hospital with symptoms which were attributed to Addison's Disease. At the autopsy the suprarenal capsules were found to be distended to the size of small oranges by extravasated blood. The liver weighed sixty-two ounces. No malignant growth was found in any organ. *See Trans. Path. Soc.* 1885, p. 454.

1304   Cirrhosis of the Liver.

A liver weighing thirty-two ounces and considerably altered in shape so as to measure ten inches transversely and four inches from back to front. Its surface is coarsely granular, and its cut section shews a white network of fibrous tissue enclosing areas of varying size filled with a softer substance. In a section examined under the microscope are seen the appearances of ordinary multilobular cirrhosis, and in addition there are collections of spheroidal cells in the branches of the portal vein suggestive of carcinomatous thrombus.

Maria S., æt. 60, was admitted under Dr. Pye-Smith with ascites of three months' duration, for the relief of which paracentesis was performed and blood-stained serum evacuated. She died five days after admission, and at the autopsy the sternum and ribs were in a condition of senile osteo-malacia. No malignant growth was found in the body. *See Insp.* 1884, No. 106.

1305   Cirrhosis of the Liver.

A portion of a liver, injected, and shewing its surface to be uniformly roughened by well-defined granular elevations. Histological examination shews the hepatic lobules to be surrounded and separated from each other by an excessive amount of fibrous tissue.

Mary P., æt. 55, was admitted under Dr. Barlow with jaundice, purpura, and œdema of the legs, and died two days later. At the autopsy three pints of ascitic fluid were found in the peritoneal cavity, and the aortic and mitral valves were thickened and contracted. The kidneys were small and granular. See Insp. vol. 37, p. 278.

1306   Cirrhosis of the Liver.

A portion of a liver presenting upon its surface numerous flattened nodules varying in size from a quarter of an inch to a line in diameter. The cut section shews the lobules of the organ to be separated from each other by an excess of fibrous tissue. Histologically there is a moderate degree of cirrhosis, the hepatic cells being unaltered.

Patrick M., a middle-aged man, was admitted under Mr. Key for a severe injury to his head, from which seven hours later he died. At the autopsy no ascitic fluid was found in the peritoneal cavity, and the membrane itself appeared quite healthy. See Insp. vol. 9, p. 155.

1307   Cirrhosis of the Liver.

A portion of a liver, the surface of which is irregularly nodulated. The cut section shews the organ to be traversed by intersecting bands of fibrous tissue. Many of the areas between the fibrous septa are engorged with blood and histologically exhibit degeneration of the hepatic cells.

George D., æt. 58, was admitted under Dr. Rees with œdema of the legs and a slight degree of ascites. The liver was felt to be enlarged and nodular. The patient had drunk alcohol freely. He died three months after admission, and at the autopsy the aorta was found to be very atheromatous, and the heart, which was dilated, weighed thirty-one ounces. See Insp. 1863, No. 252 ; and Prep. 1459 [2nd Edit.].

1308   **Cirrhosis of the Liver.**

A portion of a liver, the capsule of which is thickened and sanded over with numerous minute elevations resembling miliary tubercles. The cut surface of the organ shews an excess of fibrous tissue between the lobules. Histologically there is well-marked cirrhosis of the multilobular type, and the minute elevations above mentioned are seen to be continuous with the fibrous tissue of the capsule of the liver and to present a central lumen filled with blood-corpuscles.

From a female, æt. 23, who " was affected with dropsy."

1309   **Cirrhosis of the Liver.**

The liver of a child, the capsule of which is slightly thickened and its surface covered with rounded prominences of irregular shape and size. The cut surface shews an excess of fibrous tissue which under the microscope is seen to surround and extend into the periphery of the lobules.

From a child who died in the Evelina Hospital under the care of Dr. Goodhart in 1883.

1310   **Cirrhosis of the Liver.**

Portions of a liver the surface of which is smooth and partially covered by filamentous adhesions. The cut section shews the greater part of the organ to be replaced by fibrous tissue, in which are seen islets of yellow colour consisting of unaltered liver-cells. Histologically the cirrhosis is of the multilobular type, the fibrous tissue containing numerous ducts.

Louisa G., æt. 18, was admitted under Mr. Jacobson with cellulitis of the arm following an injury. She became delirious and comatose, and died a week after her admission. At the autopsy the liver, which weighed thirty-five ounces, was firmly adherent to the diaphragm. There was adhesive inflammation of the pleura, pericardium, and peritoneum. There was neither jaundice nor ascites nor any sign of syphilis. *See Insp.* 1889, No. 404.

1311   Cirrhosis of the Liver.

A liver, the surface of which is nodular and its capsule somewhat thickened. The cut section shows an excess of fibrous tissue between the lobules. Histologically the cirrhosis is of the multilobular type.

Edith B., æt. 14, was admitted under Dr. Taylor with ascites and dry pleurisy on the right side. She had previously on several occasions been under treatment in the hospital, paracentesis abdominis having been eleven times performed. After admission a pleuritic rub was heard on the right side, and she suffered from vomiting and epistaxis. Eleven pints of fluid were withdrawn from the abdomen, and the patient died two days after the operation. At the autopsy the liver was found to weigh 16 ounces, and the spleen, which weighed 13 ounces, presented a considerable thickening of its capsule. *See Insp.* 1894, No. 80.

1312   Cirrhosis and Capsulitis of the Liver. Thrombosis of the Portal Vein.

A portion of a cirrhotic liver mounted to shew the main trunk of the portal vein occupied by an adherent thrombus. The capsule of the liver is much thickened.

George W., æt. 33, was admitted under Dr. Pavy for ascites of two weeks' duration, having eighteen months previously been treated in the hospital for a similar condition. He died about three months after admission, a few days after the third occasion on which he had been tapped. The patient was said to have been intemperate until six years before his death. At the autopsy general peritonitis was found, and the liver, which was much deformed, weighed thirty-five ounces. *See Insp.* 1869, No. 332.

1313   Fatty and Cirrhotic Liver.

A portion of liver the cut surface of which shews many of the larger branches of the hepatic vein to be surrounded by well-defined zones of a pale yellow colour. These pale areas are seen under the microscope to be produced by extreme fatty infiltration of the liver-cells, while the rest of the organ, which presents no abnormal appearance to the naked eye, shews an excess of fibroid tissue. The surface of the liver is smooth.

Thomas H., æt. 42, was admitted under Dr. Goodhart for enlargement of the liver and spleen, with jaundice and pyrexia. There was pus in the urine. He died from facial erysipelas. At the autopsy pyelonephritis was found. *See Insp.* 1890, No. 414 ; and Drawing.

1314 Fibrosis of the Liver.

A portion of the right lobe of a liver, the cut surface of which shews it to be traversed by a thick band of white tissue invading and in parts replacing the liver-substance. Connected with this main band is a fine network of fibrous tissue, in the meshes of which are enclosed groups of liver-cells, which under the microscope are seen to have undergone fatty degeneration. The surface of the right lobe presents several rounded protuberances, that of the left lobe being coarsely granular.

George N., a publican, æt. 52, was admitted under Dr. Fagge with ascites, for the relief of which paracentesis was twice performed. He died two days after the second operation, and at the autopsy the peritoneum was found to be covered with recent lymph and the omentum was puckered as the result of chronic inflammation. The liver weighed sixty-nine ounces. The connective tissue in the portal fissure was greatly thickened, and was continuous with a dense fibrous band which traversed the substance of the right lobe of the liver. No malignant growth was discovered in the body. *See Insp.* 1883, No. 13.

1315 Circumscribed Fibrosis of the Liver.

A portion of a liver in which is seen a well-defined egg-shaped tumour two inches in length, half projecting from its upper surface. The liver itself is soft and fatty, and the tumour is of firmer consistence and traversed by bands of fibrous tissue. The histological appearances of the latter are those of well-marked cirrhosis.

1316   Calcification of the Liver.

A portion of the right lobe of a liver, the capsule of which is thickened and adherent to the diaphragm. The cut surface shews the greater part of the liver-substance to be replaced by a dense white tissue, embedded in which are orange-coloured islets of calcareous material, varying much in size and shape. Histological examination of the less diseased portion of the organ shews an excess of fibrous tissue surrounding the lobules and extending between the hepatic cells. Where the disease is more advanced no trace of normal structure remains, its place being taken by a dense fibrous tissue undergoing calcification.

From James T., æt. 62, who was affected with a hard painless enlargement of his liver for 17 years. He died from empyema on the right side after four months' illness. *See Trans. Path. Soc.* vol. 40, p. 123.

Presented by Dr. Henry Davy, 1889.

1317   Actinomycosis of the Liver.

A portion of the right lobe of a liver showing towards its anterior edge a trabeculated abscess, found on microscopical examination to contain numerous ray-fungi. The abscess projects slightly above the surface of the organ, which is roughened by old adhesions.

Esther A., æt. 36, was admitted under Dr. Horrocks with a fistulous opening at the umbilicus, through which pus had for some time been discharged. There were evidences of pelvic cellulitis, and the liver was enlarged. The patient died eight weeks after admission, and at the autopsy an abscess resembling that in the liver was found in the left ovary, and both lungs contained actinomycotic nodules. The intestines were matted together, and there was a softening thrombus in a branch of the hepatic vein. *See Insp.* 1885, No. 330.

1318   Actinomycosis of the Liver.

A portion of a liver shewing the substance of its Spigelian lobe to be for the most part replaced by a coarse network of whitish material, enclosing spaces of various sizes, which in the recent state were filled with pus containing small yellow grains. On the reverse of the specimen the disease is seen to have affected a larger area of the right lobe at its anterior part, and blue rods indicate the two sinuses through which pus was discharged from the larger abscess into the duodenum. Histological examination shews the presence of numerous actinomycetes.

Henry L., æt. 42, was admitted under Dr. Taylor complaining of pain in the right side, wasting, and night-sweats. A tumour, which was thought to be connected with the kidney, was explored through an incision in the right lumbar region, and a quantity of foul pus evacuated. The patient died ten weeks after the first onset of serious symptoms, and at the autopsy the lungs were found to be affected with actinomycosis. *See Insp.* 1890, No. 330 ; and *Guy's Hosp. Reps.* 1891, p. 312.

1319   Syphilitic Hepatitis.

A portion of a liver with smooth surface and homogeneous texture, which under the microscope is seen to be affected by cirrhosis. Columns of hepatic cells in each lobule are separated from each other by a richly nucleated fibrous stroma.

From a syphilitic infant, one month old, whose liver during life was felt to be enlarged. At the autopsy the organ was found to weigh 24 ounces, and to be almost universally affected in the manner shewn in the preparation. *See Trans. Path. Soc.* 1866, p. 167.

Presented by Dr. Wilks.

1320   Gumma of the Liver.

A portion of the right lobe of a liver, the anterior half of which is occupied by a circumscribed yellowish mass of cartilaginous consistency. The capsule covering the gumma is thickened and firmly adherent to the adjacent diaphragm. Histologically the mass consists of fibrous tissue with areas of caseous and calcareous degeneration.

Francis F., æt. 59, was admitted under Dr. Taylor with intestinal obstruction, for the relief of which a right inguinal colotomy was performed. Three days later the patient died, and at the autopsy a malignant stricture was found in the colon just below its splenic flexure. The liver weighed fifty-two ounces, the left lobe being much hypertrophied. The testes were fibrous. *See Insp.* 1892, No. 5.

1321   Gummata of the Liver.

Portions of a liver mounted to shew numerous hard yellow nodules varying in size from a line to a quarter of an inch in diameter. They are surrounded by a definite fibrous investment separating them from the hepatic tissue, which is highly lardaceous. Histologically the adventitious material is seen to be gummatous.

Thomas R., æt. 31, was admitted under Dr. Habershon for phthisis, from which 9 days later he died. At the autopsy the peritoneal cavity was found to contain 24 pints of serous fluid, and there was recent peritonitis. The liver weighed 81 ounces, and the spleen, which weighed 17 ounces, was lardaceous. *See Insp.* 1850, No. 214.

1322   Gummata of the Liver.

A section of a liver, embedded in which are several nodular masses varying in size from an eighth of an inch to an inch in diameter. They are of a yellowish colour and firm consistency, and histologically are seen to be composed of fibrous and caseous material, the walls of the arterioles in the neighbourhood being greatly thickened.

Jane W., æt. 32, was admitted under Dr. Habershon for general dropsy and albuminuria, from which one month later she died. At the autopsy the kidneys were found to be large and white, and the spleen was highly lardaceous. *See Insp.* 1864, No. 252.

1323   Softening Gummata of the Liver.

A median portion of a liver, the centre of which is occupied by a gummatous mass measuring four inches in its longest diameter. The mass presents several excavations filled with a soft bile-stained material, one of the cavities being stated in the recent condition to have been " as large as a cricket ball."

Emma W., æt. 32, was admitted under Dr. Wilks with ascites, having four months previously suffered from an attack of jaundice. Two months after admission she died from peritonitis, and at the autopsy the spleen was found to weigh 18 ounces. None of the viscera were lardaceous. *See Insp.* 1877, No. 402 ; and *Trans. Path. Soc.* vol. 29, p. 135.

1324   Gumma of the Liver obstructing the Vena Cava.

A median portion of a liver shewing beneath the thickened capsule of its convex surface a mass of gummatous deposit, which in the recent state occupied the greater part of the right lobe. The mass is made up of caseous nodules of various sizes, surrounded by fibrous tissue. The vena cava and some of the branches opening into it appear to be narrowed by contraction of the deposit.

George W., æt. 34, was admitted under Dr. Pavy for ascites and œdema of the legs. Paracentesis abdominis was twice performed, and shortly after the second operation the patient died from peritonitis. At the autopsy scars were found upon the penis and in the groins. The liver was firmly adherent to the diaphragm. *See Insp.* 1861, No. 106.

1325   Gumma and Hydatid Cyst of the Liver.

The hinder half of a liver, shewing its left lobe to be entirely occupied by a dense mass of gummatous material measuring six inches by four. In the centre of it is seen an excavation about two inches in diameter, filled with hydatid membranes. The thickened capsule of the liver is adherent to the diaphragm.

John C., æt. 39, was admitted under Dr. Goodhart for enlargement of the liver and spleen with albuminuria. He contracted syphilis 18 years previously and had suffered from partial hemiplegia. Two months after admission he died, and at the autopsy the liver was found to weigh 147 ounces, and the kidneys were lardaceous. *See Insp.* 1885, No. 160; and *Trans. Path. Soc.* vol. 37, p. 276.

1326   Congenital Syphilitic Fibrosis of the Liver.

A portion of a liver the surface of which is deformed, scarred, and covered by a thickened capsule adherent to the diaphragm. A section has been made to shew the cicatricial tissue penetrating the substance of the organ and enclosing within it small gummatous nodules. The hepatic cells are fatty and lardaceous.

Mary S., æt. 7, was admitted under Dr. Hale White with enlargement of the liver and albuminuria. Both the child's parents had suffered from syphilis, and she herself had chorciditis, and a node upon the tibia. She died from tonsillitis eight weeks after admission, and at the autopsy the liver was found to weigh 61 ounces and the spleen 16 ounces. These organs, as well as the kidneys and intestines, were lardaceous. *See Insp.* 1887, No. 235 ; and *Trans. Path. Soc.* vol. 39, p. 444.

1327   Syphilitic Fibrosis of the Liver.

A portion of a liver the capsule of which is thickened and its shape deformed by irregular depressions and fissures producing bossy elevations upon its surface. The cut sections shew the organ to be traversed by numerous white bands, which histologically are seen to consist of fibrous tissue invading the healthy liver-substance.

Charles N., æt. 23, was admitted under Dr. Goodhart with ascites of five months' duration, which had been relieved on many occasions by paracentesis abdominis. Two years previously he had had an attack of jaundice, which lasted for five months. There were well-marked evidences of congenital syphilis. The patient died suddenly on the day following his admission, and at the autopsy the peritoneal cavity was found to contain thirteen pints of turbid yellow fluid. The spleen weighed fifteen and a half ounces. *See Insp.* 1890, No. 388.

1328   Syphilitic Scarring of the Liver.

A portion of the right lobe of a liver the surface of which is marked by numerous irregular linear depressions, giving it a scarred and deformed appearance.

The capsule is thin, and the cut section shews a small yellow gumma embedded in the substance of the organ.

Henry B., æt. 23, was admitted under Dr. Pitt in a moribund condition, and died from pneumonia. At the autopsy the lower lobe of the right lung was found to be consolidated. The entire liver was marked by scars such as are seen in the preparation, and at the bottom of some of the depressions, histological examination revealed the presence of contracting gummata. *See Insp.* 1894, No. 111.

1329　Miliary Tuberculosis of the Liver.

The right margin of a liver, projecting beneath the capsule of which are seen numerous rounded bodies about the size of millet seeds, to some of which are attached tags of recent lymph. Histological examination shews the presence beneath the capsule and in the substance of the organ of miliary tubercles.

George S., æt. 10, was admitted under Dr. Addison for a cough of seven weeks' duration, and died three weeks later from general tuberculosis. At the autopsy the bronchial glands, lungs, meninges of the brain, pericardium, kidneys, and spleen were found to be affected. *See Insp.* 1858, No. 2; and *Preps.* 1445 (50), 2007 (50), & 2035 (92) [2nd Edit.].

1330　Tuberculosis of the Liver.

A portion of the right lobe of a liver presenting at its upper and hinder part a pale wedge-shaped area, the base of which, beneath the capsule, measures about three inches. The section shews the liver at this part to be invaded by a speckled yellow deposit, which under the microscope exhibits small-celled infiltration, giant cells, and caseous foci. The capsule of the organ appears to be normal.

William V., æt. 35, was admitted under Dr. Pye-Smith for phthisis of six months' duration and died four weeks after admission. At the autopsy the larynx was found to be ulcerated, and there were miliary tubercles in the kidneys. The liver was greatly enlarged, and its surface was studded all over with yellowish-white patches similar to that seen in the preparation. *See Insp.* 1876, No. 21; and *Trans. Path. Soc.* vol. 27, p. 196.

1331   Caseous Nodules in the Liver.

A portion of the right lobe of a liver, shewing two
flattened nodules projecting slightly above the capsule.
They measure about a quarter of an inch in diameter,
and the section through one of them shews that it is
embedded in the substance of the liver, from which it is
separated by a well-defined margin.   Histologically the
nodules have the characters of caseating tubercle.

Henrietta P., æt. 11 months, was admitted under Mr. Lane in
a moribund condition.   She died the next day, and at the autopsy
caseous masses were found in the lungs and bronchial glands.   *See
Insp.* 1891, No. 92.

1332   Tuberculous Abscess in the Liver.

A portion of the left lobe of a liver, the greater part of
which is occupied by caseous material.   In some parts
softening has occurred with the production of abscesses,
which are separated from each other by fibrous trabeculæ.
Below is mounted a small portion of the same lobe,
beneath the capsule of which projects an abscess the
size of a hazel-nut.   There is chronic perihepatitis.

George K., æt. 33, was admitted under Dr. Pavy for scrofulous
disease of the genito-urinary organs, symptoms of which had been
noticed for about twelve months.   He died five days after admission,
and at the autopsy the lungs, mesenteric glands, the left kidney,
the bladder, the prostate and testes were found to be affected by
tuberculosis.   *See Insp.* 1878, No. 244.

1333   Leucocythæmic Liver.

A slice of liver taken from the region of the gall-bladder
and measuring thirteen inches in its antero-posterior
diameter.   In the recent state the organ was pale, soft,
and flattened, and weighed 138 ounces.   Histologically
the hepatic tissue is seen to be uniformly infiltrated
with numerous small round cells.

From a woman, æt. 40, who was admitted under Dr. Oldham
for leucocythæmia.   The spleen was removed by operation and

found to weigh 10½ lbs. The patient died immediately after the operation, and at the autopsy more than a pint of blood was found in the region of the spleen. *See Insp.* 1867, No. 283.

1334 Liver in Pernicious Anæmia.

A portion of a liver containing an excess of iron, as shewn by the deep-blue colour imparted to it by immersion in a solution of ferrocyanide of potassium acidulated by very dilute hydrochloric acid. The percentage of iron in the dried organ was estimated by Mr. Hopkins to be 1·038, an amount more than eleven times greater than the average normal percentage. Histological examination shews that the iron is principally distributed at the periphery of the lobule.

Richard F., æt. 55, was admitted under Dr. Hale-White for well-marked pernicious anæmia of eight weeks' duration. The blood was found to contain 24 per cent. of the normal number of corpuscles and 18 per cent. of the normal quantity of hæmoglobin. The liver was slightly enlarged, and increased in size while under observation. Pericarditis supervened, and the patient died fourteen weeks after admission. At the autopsy the liver was found to weigh sixty-six ounces, and this organ as well as the kidneys and spleen contained a great excess of iron. *See Insp.* 1893, No. 224; and *Guy's Hosp. Reps.* vol. 50, p. 349, Case I.

1335 Aneurysms of the Hepatic Artery.

A portion of a liver with the structures in the portal fissure seen from the under surface of the organ. On the right branch of the hepatic artery, just at its origin, there is an aneurysmal sac measuring three and a half by two and a half inches. The sac lies upon the upper part of the gall-bladder and compresses it, and on its external surface is seen the common bile-duct, the proximal portion of which is much dilated, as are also its branches throughout the liver. Upon the left branch of the hepatic artery is a similar and smaller aneurysm embedded in the substance of the liver. The aorta is healthy.

George L., æt. 18, was admitted under Dr. Hale-White with lobar pneumonia affecting the right lung. Twenty days after admission the patient became jaundiced, and four days later pus was evacuated from the right pleural cavity. He died suddenly a month after admission, and at the autopsy the abdomen was found to contain several pints of blood-clot, the hæmorrhage having proceeded from a rent in the larger of the two aneurysms shewn in the preparation. *See Insp.* 1891, No. 117.

1336   Aneurysm of the Portal Vein.

A portion of a liver mounted to shew upon its under surface an aneurysmal sac about the size of a Tangerine orange, situated immediately to the left of the inferior vena cava and communicating by a short neck with the trunk of the portal vein. This vessel, which is seen on the reverse of the specimen, has thick walls containing calcareous material and is partially occluded by an organised thrombus extending into both its collecting and distributing branches. The capsule of the liver is thickened, and the hepatic tissue is seen under the microscope to be very fatty.

Cordelia W., æt. 34, was admitted under Dr. Habershon for ascites of eight months' duration, for which paracentesis was on several occasions performed. She died six weeks after admission, twelve days after the last occasion on which she was tapped. At the autopsy the peritoneal cavity was found to contain a considerable quantity of lymph and recent blood-clot. The spleen was much enlarged and presented numerous infarcts. *See Insp.* 1870, No. 265.

1337   Thrombosis of the Portal Vein. Cirrhosis of the Liver.

A portion of a liver mounted to shew the main trunk of the portal vein occupied by a firm adherent thrombus. Histological examination of the liver shews a slight excess of fibrous tissue in Glisson's capsule.

Thomas L., æt. 32, was admitted under Dr. Taylor with an enlarged spleen, and died two months later from profuse hæmatemesis. At the autopsy the thrombus in the portal vein was found to extend into the splenic and superior mesenteric veins, which were greatly dilated and tortuous. The spleen was enlarged to forty-eight ounces, and its enlargement was thought to be partly due to former attacks of ague. The peritoneal cavity contained much ascitic fluid. *See Insp.* 1881, No. 313.

1338   Persistent Umbilical Vein.   Cirrhosis.

A portion of a liver shewing in the position of the round ligament an injected vessel, which is the persistent umbilical vein. This vein communicates freely with the trunk of the portal vein. The surface of the liver is "hobnailed," and microscopically there is a slight degree of cirrhosis.

Thomas E., æt. 53, was admitted under Dr. Taylor in a drowsy condition, with twitching of the limbs and albuminuria. He died six weeks after his admission. He had been "a great drinker," and ten years previously was treated in the hospital for hæmatemesis and cirrhosis of the liver. At the autopsy it was found that water injected into the portal trunk flowed freely through the persistent umbilical vein. The kidneys were contracted and granular. *See Insp.* 1884, No. 416.

1339   Cavernous Angeioma of the Liver.

A portion of a liver shewing beneath its capsule a mass of spongy tissue, measuring two inches in length and extending one inch into the substance of the organ. On the surface the mass is seen to have a well-defined margin and to be level with the surrounding parts. In the recent state it was of a dark red colour, and histologically it consists of cavernous tissue filled with blood-corpuscles.

Mary W., æt. 74, was admitted under Mr. Birkett with a strangulated femoral hernia. The patient had a mole on the chin and a large nævus on the forehead. *See Insp.* 1863, No. 247.

1340   Cavernous Angeioma of the Liver.

A portion of a liver shewing beneath its capsule an oval mass of spongy tissue, measuring an inch in its longest diameter. The mass is circumscribed and in the recent state its meshes were filled with dark blood.

D 2

Jemima C., æt. 63, was admitted under Mr. Bryant with elephantiasis of the left lower extremity, and died a few days after amputation through the thigh. At the autopsy the liver was found to contain two other large and several small angeiomata. *See Insp.* 1874, No. 417.

1341   Adenoma of the Liver.

A marginal portion taken from the left lobe of a liver, containing an encapsuled globular mass measuring an inch and a quarter in diameter and protruding slightly from both surfaces of the organ. The tumour does not differ in appearance from normal liver-substance, except for the presence of an excess of fibrous tissue between the component lobules, and under the microscope its characters are those of multilobular cirrhosis.

James B., æt. 26, was admitted under Mr. Bransby Cooper in 1827 with a strangulated hernia, for relief of which herniotomy was performed. Two days later he died and at the autopsy the liver was found to be healthy, except for the tumour which forms the preparation. *See Insp.* vol. 4, p. 37 ; and *Prep.* 1115.

1342   Cylindrical-celled Carcinoma of the Liver.

A portion of liver in the hinder part of which is embedded a circumscribed mass, oval on section, and measuring two and a half inches in its longest diameter. The growth, which is white and traversed by a fine network of fibrous trabeculæ, has the histological characters of a cylindrical-celled carcinoma.

From a patient who died from persistent diarrhœa. At the autopsy similar deposits were found in the mesenteric glands, some of which had softened and communicated with the intestine. *See Wax Model,* No. 71.

Presented by Mr. Hilton.

1343   Cylindrical-celled Carcinoma of the Liver.

A portion of a liver the hinder part of which is occupied by an infiltrating deposit of white growth, circumscribed nodules of a similar deposit being scattered through the rest of the organ. Beneath the capsule some of these nodules appear as flattened prominences, having shallow central depressions. Histologically the growth consists of fibrous tissue in which are small alveoli lined with cubical epithelium and many of them filled with spheroidal cells.

Alfred W., æt. 48, was admitted under Dr. Pavy with a smooth and painful enlargement of the liver. He gradually became paraplegic, and died from bronchitis. At the autopsy the liver, which was thought to be the primary seat of the growth, was found to weigh 120 ounces, and there were secondary deposits in the spine, lungs, spleen, kidneys, and lymphatic glands. See Insp. 1887, No. 2; and Trans. Path. Soc. vol. 38, p. 151.

1344   **Cylindrical-celled Carcinoma of the Liver.**

A portion of a liver, shewing a wedge-shaped deposit of firm white growth at the posterior part of the organ close to the inferior vena cava. Histologically the growth is a cylindrical-celled carcinoma.

Ellen G., æt. 36, was admitted under Dr. Taylor with an artificial anus formed eleven months previously for the relief of intestinal obstruction. An attempt was made to close the fistula, and the patient died from peritonitis. At the autopsy a malignant stricture was found in the descending colon, and there were several secondary deposits of growth in the liver. See Insp. 1893, No. 220.

1345   **Carcinoma of the Liver.**

A portion of a liver mounted to shew on its surface numerous irregular depressions, over which the capsule is thickened and roughened by adhesions. Corresponding to these depressions are deposits of a scirrhous growth, which is seen on the cut surface to infiltrate and in a great measure to replace the substance of the liver. Histologically the growth is a spheroidal-celled carcinoma with much stroma.

Maria O., æt. 50, was admitted under Dr. Goodhart with an enlarged liver and a contracting growth in her right breast, the

nipple of which had been drawn in for nine years. She died
eleven days after admission, and at the autopsy the lungs were
found to be puckered by the contraction of secondary deposits.
See Insp. 1891, No. 18 ; and *Prep.* A 3351.

1346   Spheroidal-celled Carcinoma of the Liver.

A slice taken from the right lobe of a liver the cut
surface of which shews the organ to be extensively
infiltrated by a white deposit, which histologically has
the characters of a spheroidal-celled carcinoma with
scanty stroma. The capsule is in parts thickened, and
presents irregular depressions.

> Elizabeth M., æt. 38, was admitted under Dr. Taylor with a
> painful and distended abdomen. She died two weeks after admis-
> sion, and at the autopsy fifty ounces of fluid were found in the
> peritoneal cavity. The liver, which was uniformly infiltrated with
> growth, weighed seventy ounces. In the right breast there was a
> hard mass of carcinoma an inch and a half in diameter. There
> were no other secondary deposits. *See Insp.* 1894, No. 10.

1347   Carcinoma of the Liver.

A portion of a liver shewing the anterior part of its
right lobe to be occupied by an oval mass of growth
measuring four inches in its longer diameter and three
inches in thickness. The central part of the growth is
excavated, and its surface is covered by a rough capsule
of fibrous tissue, to which portions of the duodenum and
transverse colon are firmly adherent. Histologically the
growth is a spheroidal-celled carcinoma with abundant
stroma.

1348   Contracting Carcinoma of the Liver.

A portion of a liver the surface of which is broken up
by numerous deep irregular depressions in which the
capsule is opaque and thickened. The cut surface shews
these depressions to be due to the contraction of a firm
white growth infiltrating and in great part replacing

the hepatic tissue. Histologically the growth is a scirrhous spheroidal-celled carcinoma, the central parts of which are composed almost entirely of fibrous tissue. The liver is extremely fatty.

Emma F., æt. 45, was admitted under Dr. Wilks for a hard tumour in the right hypochondriac region associated with ascites. Her abdominal symptoms were of three months' duration. She died five weeks after admission, and at the autopsy carcinomatous growths were found in the breast, ovaries, peritoneum, pleura, and pericardium. See Insp. 1871, No. 49.

1349   Scirrhous Carcinoma of the Liver.

A portion of a liver infiltrated by a firm white deposit, which histologically has the characters of a spheroidal-celled carcinoma with much stroma. The capsule of the organ is thickened, and its surface irregular and nodulated.

Robert M., æt. 64, was admitted under Dr. Rees for enlargement of the liver with ascites, the latter symptom having been present for about three weeks. He died a week after admission, and at the autopsy secondary deposits were found in the lung and in the glands of the portal fissure. The colon contained about a dozen polypoid growths which were thought to be secondary to the carcinoma of the liver. See Insp. 1868, No. 149.

1350   Infiltrating Carcinoma of the Liver.

A portion of a liver the lower two-thirds of which are occupied by a homogeneous white growth. The surface of the organ is of normal appearance, except that the capsule over the affected part is slightly roughened by a deposit of lymph. Histologically the growth is a spheroidal-celled carcinoma with much stroma.

Anne B., æt. 44, was admitted under Mr. Bransby Cooper in 1830 for cancer of the uterus, the symptoms of which had been noticed for four months. She died twelve weeks after admission, and at the autopsy the uterus and the posterior wall of the bladder were found to be partially destroyed by malignant ulceration, and there were secondary deposits in the lumbar glands. See Insp. vol. 10, p. 93.

1351   Colloid Carcinoma of the Liver.

A portion of a liver the greater part of which is occupied by a circumscribed mass of gelatinous growth presenting a central excavation. The mass projects beneath the capsule of the organ as a prominent rounded swelling, the surface of which is in parts wrinkled from hardening and contraction by spirit. Histologically the structure is that of a spheroidal-celled carcinoma, many of the cells of which have undergone colloid change.

Eleanor M., æt. 32, was admitted under Dr. Barlow for vomiting and anæmia. On admission there was a prominent abdominal tumour, which had been noticed for "a long time." At the autopsy the wall of the stomach was found to be infiltrated by a growth which had extended into the anterior abdominal parietes. *See Insp.* 1865, No. 109; and *Prep.* 696.

1352   Melanotic Carcinoma of the Liver.

A portion of a liver shewing beneath its capsule a black mass, ovoid in shape and measuring half an inch in its longest diameter. The nodule has a well-defined capsule and projects but slightly above the surface of the organ. Histologically the growth is a spheroidal-celled carcinoma, the cells of which contain much black pigment.

From Mrs. F., who was under Mr. Birkett's care for a tumour of the breast, which had been noticed for two years before her death. At the autopsy malignant growth was found in the breast, in the lungs, and in the mediastinal glands. In none of these situations was the growth observed to be melanotic.

1353   Calcifying Carcinoma of the Liver.

A portion of a liver injected and incised to shew its central parts occupied by a hard fibrous mass, with numerous radiating processes traversing the substance of the organ. In the central mass are seen several islets of calcareous material. The surface of the liver presents many rounded prominences due to malignant deposit. Histologically the growth is a spheroidal-celled carcinoma with abundant stroma.

William G., æt. 21, was admitted under Dr. Cholmeley in 1828 for ascites, and died three weeks after admission. At the autopsy the peritoneum was found to be thickly studded with secondary deposits. *See Insp.* vol. 6, p. 109 ; and *Prep.* 1227.

1354 Carcinomatous Excavations in the Liver.

A portion of a liver studded with numerous masses of growth of various sizes. Two of the larger masses present excavations having ragged walls and containing partially detached fragments of disintegrating growth. Histologically the nodules have the characters of spheroidal-celled carcinoma with a considerable amount of stroma.

Robert W., æt. 32, was admitted under Dr. Barlow with jaundice, ascites, and œdema of the legs, and died a month after admission. At the autopsy the liver was found to weigh 9½ lbs., and was almost entirely occupied by growth which had extended through the diaphragm into the right pleura. There was carcinomatous ulceration of the sigmoid flexure, with secondary deposits in the lumbar glands. *See Insp.* 1855, No. 183.

1355 Carcinomatous Thrombus of the Portal Vein.

A portion of a liver mounted to shew the main trunk of the portal vein distended by a firm cancerous thrombus, the distal part of which is of about the size and shape of a pigeon's egg and is not adherent to the vessel-wall. Within the organ are seen several deposits of secondary growth, which histologically have the characters of spheroidal-celled carcinoma.

John S., æt. 30, was admitted under Dr. Pavy for vomiting and diarrhœa, with occasional hematemesis and melæna. The liver was felt to be enlarged and nodular. He died two months after admission, and at the autopsy the walls of the stomach were found to be uniformly infiltrated with malignant growth, and there were numerous secondary deposits on all parts of the peritoneum. *See Insp.* 1873, No. 393.

1356   Carcinoma of the Liver invading the Vena Cava.

A portion of a liver mounted to shew a mass of growth invading and partially occluding the inferior vena cava. On the cut surfaces are seen numerous masses of white growth, which have the histological characters of cylindrical-celled carcinoma.

Charles M., æt. 67, was admitted under Dr. Taylor with ascites and œdema of the legs, from which he had suffered for four months. Paracentesis abdominis was performed, and the patient died ten days after his admission. At the autopsy malignant growth was found in the stomach, and there were secondary deposits in the lung, pleura, and bronchial glands. *See Insp.* 1891, No. 124.

1357   Cirrhosis and Carcinoma of the Liver.

An antero-posterior slice from the right lobe of a liver, shewing at its upper part its substance replaced by numerous closely aggregated nodules of growth embedded in a fibrous stroma. The rest of the hepatic tissue shews the lobules to be separated from each other by an excessive thickening of connective tissue. Histologically the growth is a spheroidal-celled carcinoma, and the liver is cirrhotic.

John R., æt. 49, was admitted under Dr. Goodhart for ascites of two months' duration. He had been a heavy drinker for fifteen or sixteen years, but had given up drink six years before his death. Blood-stained serum was withdrawn from the peritoneal cavity and from the right pleura. The patient died three weeks after his admission, and at the autopsy secondary deposits were found in the bronchial and portal glands, and it was thought that the liver was the primary seat of the growth. *See Insp.* 1892, No. 410.

1358   Carcinoma and Cirrhosis of the Liver.

A portion of a liver the cut surface of which shews a great excess of fibrous tissue replacing the hepatic substance. A main branch and many of the smaller branches of the portal vein are filled with thrombus, which under the microscope is seen to consist of large spheroidal cells embedded in a scanty stroma. The liver-substance is cirrhotic, and in the fibroid tissue are observed alveoli of various sizes filled with cells similar to those forming the thrombus.

Isaac B., æt. 68, was admitted under Dr. Pye-Smith with an enlarged liver, which was painful and nodular. Ascites supervened, and the patient died, comatose, sixteen days after admission. At the autopsy a healed ulcer was found upon the lesser curvature of the stomach. The liver weighed 71 ounces, and was the only organ in which malignant deposit was discovered. *See Insp.* 1891, No. 273.

1359   **Sarcoma of the Liver.**

A portion of a liver shewing a mass of growth about the size and shape of a pigeon's egg, situated immediately beneath the capsule and projecting about one-third of an inch above the surrounding surface. The prominence shews marked umbilication. There is a second smaller nodule embedded in the substance of the organ. Histologically the growth is a small round-celled sarcoma.

John F., æt. 30, was admitted under Mr. Key in 1821 for paraplegia. At the autopsy growths were found in the spine, pleura, pericardium, and spleen. *See Prep.* 306; and *Preps.* 1028, 1042, 1449, 1544, 1548, & 2012 (2nd Edit.).

1360   **Sarcoma of the Liver.**

A marginal portion of liver the surface of which presents numerous rounded prominences, some of which are cupped. The cut section shews the prominences to be due to masses of a soft white growth, which histologically has the structure of small round-celled sarcoma.

Hetty II., æt. 38, was admitted under Dr. Fagge for cough and dyspnœa of a few months' duration, and died six weeks after admission. At the autopsy deposits of malignant growth were found in the mediastinal and cervical glands, and in the pericardium, ovaries, and mucous membrane of the intestine. The liver weighed 170 ounces. *See Insp.* 1881, No. 240.

1361   Sarcoma of the Liver.

A portion of a liver irregularly fissured and deformed, and shewing its capsule in many parts thickened and roughened by adhesions. The cut surface shews the anterior two-thirds of the organ to be infiltrated by a firm white deposit, which histologically has the structure of small round-celled sarcoma.

Arthur A., æt. 32, was admitted under Dr. Wilks with an enlarged liver and general swelling of the lymphatic glands, the latter condition having been noticed for about six months. During his illness he suffered from paralysis of the third nerve, from which he recovered during the administration of potassium iodide. He died ten weeks after admission, and at the autopsy the liver was found to weigh 88 ounces, the middle third of the organ being affected in the manner shewn in the preparation. There were patches of fibroid tissue in the heart muscle, and the lymphatic glands in the pharynx, portal fissure, and mesentery were enlarged, firm, and succulent. The spleen weighed 6 ounces, and contained one small nodule of white deposit. *See Insp.* 1883, No. 299.

1362   Sarcoma of the Liver.

Portions of a liver mounted to shew numerous soft white nodules varying in size from a swan-shot to a cherry, some of which are deeply embedded in the organ, whilst others project beneath the capsule. Histologically the deposit consists of circumscribed nodules of small round cells. The liver is fatty.

Presented by Mr. Lawford Knaggs, 1888.

1363   Melanotic Sarcoma of the Liver.

A portion of a liver the cut surface of which shews it to be thickly beset with soft black nodules varying in size from a millet-seed to a walnut. Histologically the structure is that of a spindle-celled sarcoma containing much pigment.

From a patient who suffered from a melanotic growth of the face. *See Wax Models,* Nos. 72 & 290.

1364   Melanotic Sarcoma of the Liver.

A section of a liver largely occupied by masses of black growth, some of which project beneath the capsule and are umbilicated. Histologically the growth is a pigmented sarcoma, most of the cells of which are oval in shape.

From a woman who suffered from a melanotic tumour affecting the eyeball and the side of the face. *See Prep.* 1668 (64) [2nd Edit.]; and *Wax Model*, No. 37.

1365   Melanotic Sarcoma of the Liver.

A section of a liver studded with nodules of melanotic growth, which histologically has the structure of round-celled sarcoma, many of the cells being loaded with black pigment.

From a Greek whose liver was found to weigh 15 lbs., and in whose skin and pancreas similar growths were present.

From the Fort Pitt Museum.

1366   Melanotic Growth of the Liver.

Two sections of a liver, one of which is almost universally infiltrated with melanotic growth, whilst in the other the black deposit occurs in the form of scattered nodules. The capsule of the organ over the affected part is thickened and was adherent to the diaphragm. Histologically the growth has been regarded by some as a sarcoma, by others as a carcinoma.

George C., æt. 66, was admitted under Dr. Pavy for an abdominal tumour, associated with acute epigastric pain. His right eye had been removed for "glaucoma." He died six weeks after admission, and at the autopsy the liver was found to weigh 122 ounces, no growth being discovered in any other organ of the body. *See Insp.* 1885, No. 362; and *Trans. Path. Soc.* 1886, p. 272.

1367   Myeloid Sarcoma of the Liver.

A portion of a liver in which is seen beneath the capsule

a circumscribed mass of growth discoloured by extravasated blood. Histologically the growth is a very vascular
sarcoma, with round, oval, and large multinucleated
cells.

Susan G., æt. 45, was admitted under Dr. Pavy with rapidly
growing lumps beneath the skin and bleeding from the gums.
Her illness, which had lasted three weeks before admission, ended
fatally a fortnight later with cerebral symptoms. At the autopsy
similar growths to those in the liver were found in the lungs,
spleen, and brain. See Insp. 1861, No. 62; and Prep. 333.

1368   Hæmorrhagic Sarcoma of the Liver.

A section of a greatly enlarged liver, the substance of
which is almost entirely replaced by a dense fibrous
growth, mottled with numerous hæmorrhages. The
inferior vena cava is distended by an adherent thrombus.
Histologically the growth is a spindle-celled sarcoma.

1370   Softening Sarcoma of the Liver.

A portion of liver which has been partially injected and
presents on its cut surface numerous circumscribed
masses of soft growth varying from one eighth of an
inch to an inch and a quarter in diameter. All but the
smallest nodules have undergone softening and exhibit
excavations which cause the liver to appear to be riddled
by abscesses. Histological examination of the smaller
nodules shews them to have the structure of round-celled
sarcoma.

From Edward S., æt. 55, who became emaciated, and suffered
from ascites and anasarca a few weeks before his death. At the
autopsy the liver was found to weigh 13¾ pounds, and no malignant growth was discovered in any other part of the body. See
Note Book, p. 42.

Presented by Mr. J. T. Lipscomb, 1838.

1371   Lymphadenoma of the Liver.

A portion of a liver enlarged by lymphadenomatous

deposit, which in some places appears beneath the capsule
in the form of nodules of various sizes, over which the
peritoneum is thickened and covered by small tags of
fibrous tissue. Elsewhere the infiltration is uniform and
the capsule is smooth and of normal appearance. Histologically the deposit consists of small round cells, which
in some parts almost entirely replace the liver-substance.
In less affected parts there is a small-celled infiltration
in Glisson's capsule and between the columns of cells
forming the lobules.

Charles D., æt. 52, was admitted under Dr. Pye-Smith with
swollen glands in the neck, axillæ, and groins. The liver and
spleen were felt to be enlarged. The temperature was raised and
there was œdema of the legs, with effusion into both pleural cavities.
The patient died seven months after the hepatic enlargement was
first noticed. At the autopsy the liver was found to weigh 141
ounces, and there were numerous lymphadenomatous deposits in the
glands throughout the body. The spleen, which weighed 35
ounces, contained similar deposits. *See Insp.* 1883, No. 322.

1372   Cystic Disease of the Liver.

A portion of a liver the greater part of which is converted into a congeries of thin-walled cysts varying in
size from a millet-seed to a walnut. Histological examination shews them to have a fibrous wall and to be
lined by a layer of short columnar epithelium.

From a patient whose right kidney presented a similar cystic
condition. *See Prep.* 2047 (75) (2nd Edit.) ; and *Trans. Path. Soc.*
vol. 7, p. 235.

Presented by Mr. Key.

1373   Retention Cyst of the Liver.

A portion of the left lobe of a liver, incised to shew
immediately beneath the capsule a cyst which is about
the size of a walnut and in the recent state contained a
colourless opalescent fluid. The wall of the cyst, which
presents numerous lacunæ, is thin, and histologically is
found to be composed of fibrous tissue having no epithelial covering.

John G., æt. 62, was admitted under Mr. Jacobson with a sarcoma of the femur, for which amputation was performed. The patient died a month after the operation, and at the autopsy secondary deposits were found in the ileum and lungs. The liver weighed fifty-one ounces, and appeared normal except for the cyst shewn in the preparation. *See Insp.* 1894, No. 170.

1374   Multiple Hydatids of the Liver.

The liver of a child, containing numerous hydatid cysts, some immediately beneath the capsule, others more deeply embedded in the organ. On section the cavities are seen to vary from an inch to a third of an inch in diameter and to be lined with a delicate hydatid membrane.

From a child who died in the Evelina Hospital in 1888. The lungs were similarly affected.

Presented by Dr. Goodhart.

1375   Hydatid of the Liver.

A portion of a liver shewing at its hinder part, close to the inferior vena cava, a globular cyst measuring two inches in diameter. Its fibrous wall, which projects slightly above the surface of the organ, is partially encrusted with calcareous material.

Jane M., æt. 44, was admitted under Dr. Addison in 1835 for phthisis, from which she died. At the autopsy the cyst in the liver was found to contain hydatid membrane. *See Insp.* vol. 21, p. 51.

1376   Cirrhosis and Hydatid of the Liver.

A portion of a cirrhotic liver mounted to show at its anterior margin the fibrous capsule of a hydatid cyst, oval in shape, and measuring two inches in its longest diameter. Below is mounted a mushroom-like mass of laminated fibrous tissue, which in the recent state was attached to the capsule of the liver over the situation of the cyst.

John W., æt. 65, who had drunk spirits freely, was admitted under Dr. Habershon for jaundice and ascites. Two days later

he died, and at the autopsy the liver was found to weigh 58 ounces.
The stomach and intestines contained much blood. The cyst con-
tained hydatid membrane, with bile-stained fluid and small calculi.
See Insp. 1870, No. 161½.

1377 Hydatid Cyst of the Liver.

The right lobe of a liver mounted to shew at its hinder
part a large hydatid cyst, measuring about eight inches
transversely and projecting from the posterior surface
of the organ. The anterior edge has been pushed downwards. Upon its upper surface, six inches from its
anterior border, is seen a track, indicated by a blue rod,
passing through the substance of the liver, communicating with the portal vein and the interior of the cyst.

Edward S., æt. 40, was admitted under Mr. Bryant with a
firm elastic swelling in the right hypochondrium, which extended
a little below the umbilicus, and had been noticed for about eleven
months. A trocar was introduced three and a half inches to the
right of the middle line close to the edge of the ribs. After it had
passed through an inch and a half of the liver some blood escaped.
The instrument was pushed further and about 9 ounces of clear
fluid were evacuated through the cannula. A few moments afterwards the patient became livid, vomited two or three times, had
an epileptiform seizure, and died in twenty-five minutes. No anæs-
thetic had been administered. At the autopsy it was found that
the hydatid cyst above described contained 3½ pints of blood-stained
fluid, and that it had pushed forward the structures in the portal
fissure. The trocar had entered the trunk of the portal vein
through the convex surface of the organ. It was thought that
death was due to the escape of the hydatid fluid into this vessel.
There was no ascites. *See Insp.* 1878, No. 35.

1378 Hydatid Cyst of the Liver.

A portion of the wall of a hydatid cyst from a liver,
presenting upon its inner surface a cluster of warty
excrescences, which histologically are found to consist
of convoluted laminated membrane and caseous débris.

1379 Hydatid Cyst of the Liver.

A portion of a liver in which is seen a cyst about two

and a half inches in diameter, having a fibrous wall and containing hydatid membranes and a quantity of inspissated bile.

1380   Hydatid of the Liver.

A portion of a liver mounted to shew a thick fibrous-walled sac about two inches in diameter, which in the recent state was filled with "a thick mortar-like substance containing hydatid cysts."

Anne C., æt. 46, was admitted with hæmatemesis, jaundice, and ascites, from which she died. At the autopsy the liver was found to be cirrhotic and scarred. *See Insp.* vol. 12, p. 45, and *Prep.* 2261 (20) (2nd Edit.).

1381   Hydatid Cysts of the Liver and Peritoneum.

Three large cysts which were removed from the liver and peritoneal cavity. Portions of the wall of the cysts have been removed to expose within them numerous daughter cysts similar to those seen at the bottom of the preparation-glass.

Edward C. was admitted under Dr. Cholmeley in 1826 with a distended abdomen in which fluctuation could be felt. The abdominal wall was resistant to the touch, and gave a sensation of irregularity to the hand passed over it. At the autopsy the peritoneal cavity was found to be filled with hydatid cysts of various sizes, and there were similar cysts in the spleen. *See Red Insp. Book*, p. 170.

1382   Hydatid of the Liver.

A portion of the right lobe of a liver shewing at its posterior part a hydatid cyst measuring two inches in diameter, the walls of which are composed of fibro-calcareous material. The contents of the cyst are formed of a putty-like mass, embedded in which are seen pieces of hydatid membrane. Histological examination shews this mass to consist of granular débris, fat, and cholesterine.

Daniel H., æt. 61, was admitted under Mr. Davies-Colley with cerebral symptoms attributed to injury. He was trephined, and died comatose on the eighth day after the operation. At the autopsy the cerebral vessels were found to be very atheromatous, and there were numerous patches of softening in the brain. *See Insp.* 1890, No. 232.

1383   Old Hydatid of the Liver.

A portion of the right lobe of a liver shewing beneath its capsule a cyst measuring three quarters of an inch in diameter, the walls of which are formed of fibrous and calcareous material. Its contents consist of hydatid membranes and cretaceous débris. Over the cyst the capsule of the liver is depressed and presents a puckered appearance.

John H., æt. 47, was admitted under Dr. Rees with hemi-plegia of the right side, and died eight days after admission. At the autopsy the pons Varolii was found to be softened and the cerebral arteries were obstructed. *See Insp.* 1869, No. 146.

1384   Hydatid Cyst opening into the Hepatic Duct.

A portion of a liver seen from behind and shewing in the region of the portal fissure a hydatid cyst, which in the recent state contained two or three pints of pus. The wall of the sac has been in part removed to shew an opening, indicated by a yellow rod, which leads directly into the main hepatic duct. Within the cavity is seen a detached hydatid membrane which has passed through this opening into the hepatic and common bile-ducts and projects through the papilla into the duodenum.

Hiram B., æt. 50, was admitted under Dr. Barlow with an enlarged and tender liver and slight jaundice, the latter symptom having been present for about a month. He died a week after admission, and at the autopsy the intestines were found to be matted together by recent lymph, and the liver contained a large suppurating cyst adherent to the abdominal wall. There were two smaller cysts, one of which was suppurating. The hepatic ducts throughout the liver were filled with pus. *See Insp.* 1816, No. 13.

E 2

1385   Hydatid Membranes.

A large quantity of hydatid membrane which was removed from the liver and pleural cavity. Histologically it is found to present a laminated structure.

From James S., who, having suffered for about three years from an enlargement of the liver, was suddenly seized with dyspnœa, and on percussion was found to have dulness affecting the right side of the chest. Paracentesis was performed several times, and large quantities (on one occasion as much as 11 pints) of serous fluid were evacuated. He died two months after the onset of acute symptoms, and at the autopsy a large cyst was discovered in the liver, which communicated, through an opening in the diaphragm about the size of a crown-piece, with the right pleural cavity. The lung was much compressed.

Presented by Dr. Gull.

1386   Hydatid Cysts passed per Rectum.

A large number of thin-walled hydatid cysts of various sizes which were evacuated per anum.

From a woman æt. 34, who for about fifteen months had suffered from sickness and discomfort after food. For eight months she had experienced pain in the region of the gall-bladder, and for three months had been slightly jaundiced. She then began to pass hydatids by stool, and this continued for some time. The abdomen became softer and less tender, and the patient made a slow recovery.

Presented by Dr. Hooper.

1387   Calcified Parasites in the Liver.

A portion of a liver beneath the capsule of which is seen a cluster of seven calcareous nodules, which are thought to be the remains of an encysted parasite, probably *Pentastoma denticulatum*.

Presented by Dr. Fagge, 1879.

Section XXVI.—INJURIES AND DISEASES OF THE GALL-BLADDER.

1388   **Laceration of the Gall-Bladder.**

A gall-bladder laid open from behind to shew upon its anterior wall a transverse rent three quarters of an inch in length. On the reverse of the specimen the serous surface is seen to be coated with a thick layer of bile-stained lymph.

Daniel J., æt. 29, was admitted under Mr. Jacobson, having 24 hours previously been kicked in the abdomen by a man. Symptoms of peritonitis supervened, and 17 days later the patient died. At the autopsy the peritoneal cavity was found to contain much bile-stained fluid, and the intestines were matted together by effused lymph. There was a laceration of the liver, two inches in length, close to the gall-bladder. *See Insp.* 1881, No. 253.

1389   Gall - Bladder and Bile - Ducts distended by Blood. Cholecystotomy.

A gall-bladder with the cystic, hepatic, and common bile-ducts, together with the adjacent portions of the liver and anterior abdominal wall. The gall-bladder has been laid open and shews a longitudinal laceration about two and a half inches in length on its anterior wall towards the right side. The rent extends through the coats of the bladder and its edges are covered by adherent blood-clot. The cystic and the lower part of the common bile-duct are slightly dilated, the rest of the latter and the hepatic ducts being enormously distended. The fundus of the gall-bladder has been sutured to the abdominal wall and has been opened a little to the left of its long axis. The drainage-tube inserted at the operation remains in situ. Below are mounted the blood-clots removed at the autopsy from the gall-bladder and the hepatic ducts, the former measuring two and a half and the latter one and a quarter inches transversely.

From Mrs. L., æt. 54, who, while suffering from jaundice of two months' duration, was suddenly seized with acute abdominal pain and collapse associated with a rapidly-increasing tumour of the gall-bladder. A considerable quantity of blood was passed per rectum. Cholecystotomy was performed by Mr. Lane five days after the onset of acute symptoms, and nearly a pint of blood-clot removed from the bladder. The patient died a few hours after the operation. *See Clin. Soc. Reps.* 1895, p. 160.

Presented by Mr. Reginald Clarke.

1390   Congenital Absence of the Gall-Bladder.

The liver of an infant from which the gall-bladder and cystic duct are absent.

From an infant, æt. 10 months, who died in 1827 from convulsions. *See Insp.* vol. 3, p. 68.

1391 **Deformity and Malposition of the Gall-Bladder.**

A portion of the right lobe of a liver mounted to shew the gall-bladder dilated and turned to the left, so that its greater part forms an egg-shaped tumour three inches in length, lying parallel to and projecting beyond the anterior edge of the organ. Running upwards and to the right from the notch for the gall-bladder there is a deep furrow on the anterior surface of the liver.

John W., æt. 58, was admitted under Dr. Perry for persistent vomiting, and died from carcinoma of the stomach. At the autopsy a malignant stricture of the pylorus was discovered, and there were numerous filiform polypi in the ileum. *See Insp.* 1894, No. 363.

1392 **Dilated Gall-Bladder.**

A greatly dilated gall-bladder, which measures eight inches in length and six inches in circumference at its widest part. The obstruction which led to the distension was caused by cancerous disease at the head of the pancreas.

1393 **Distension of the Gall-Bladder.**

The anterior half of a liver, mounted to shew the gall-bladder projecting as a pear-shaped sac four inches beyond the margin of the organ. The liver is small, and its capsule is wrinkled and presents a slight furrow with local thickening thought to be due to the pressure of a belt.

John II., æt. 81, was admitted under Mr. Durham with a fracture of the neck of the left femur, from the effects of which he died ve weeks later. At the autopsy the liver was found to weigh thirty-nine ounces, and the bile-ducts were free from obstruction. *See Insp.* 1891, No. 418.

1394 **Dilatation of the Gall-Bladder and Bile-Ducts.**

A portion of a liver with the gall-bladder. The latter is dilated and in the recent state contained forty-two ounces of very slightly bile-stained fluid. The hepatic duct would admit the thumb, and the cystic duct the

little finger.   The ducts throughout the liver are seen
to be considerably dilated.

Elizabeth H., æt. 25, was admitted under Dr. Habershon for
jaundice of six months' duration associated with enlargement of
the liver and gall-bladder.   She became emaciated, and died
16 weeks after admission.   At the autopsy the liver was found to
weigh 110 ounces, and the head of the pancreas was occupied by
a scirrhous carcinoma which invaded and obstructed the common
bile-duct.   *See Insp.* 1880, No. 103.

**1395   Distended   and   Inflamed   Gall - Bladder   in
Enterica.**

A portion of a liver with the gall-bladder mounted to
shew the latter greatly distended, so as to measure in
the recent state seven and a half inches in length.
Its wall is thin, and on its serous surface it presents
a deposit of granular lymph.

Henry A., æt. 19, was admitted under Dr. Hale-White during
a relapse of typhoid fever, which proved fatal at the end of
the seventh week.   He suffered from profuse diarrhœa, and, for
the two days preceding his death, from frequent vomiting.   At
the autopsy the gall-bladder was found to contain fifteen ounces of
watery bile, which, upon slight pressure, flowed freely into the
duodenum.   *See Insp.* 1890, No. 435.

1396   Suppurating Gall-Bladder.

A portion of the right lobe of a liver, shewing the gall-
bladder to be replaced by a mass of inflammatory tissue
to which the omentum was adherent.   In the midst of
this mass is a cavity from which a yellow rod has been
passed along a sinus into the duodenum, which it enters
about an inch from the pylorus.   On the reverse of the
specimen is seen a large hydatid cyst projecting from
the under surface of the liver.

Matilda M., æt. 69, was admitted under Dr. Perry with the
signs of acute peritonitis.   She had suffered for the previous three
months from occasional attacks of abdominal pain and vomiting.
She died twelve hours after admission, and at the autopsy an
abscess was found in the region of the gall-bladder communicating
with the peritoneal cavity, which contained thirty ounces of
yellowish purulent fluid.   *See Insp.* 1892, No. 272.

1397 Gangrene of the Gall-Bladder.

The anterior part of a gall-bladder, the mucous surface of which is of a dark brown colour. On the reverse of the specimen the serous coat presents numerous defined patches of gangrene similar in colour to the mucous membrane.

Rosalind H., æt. 41, was admitted under Dr. Moxon for jaundice of 3 months' duration. A pelvic tumour was discovered and the liver was noticed to be enlarged. She died one month after admission, and at the autopsy both ovaries, the lungs, and the liver were found to be affected with carcinoma. The gall-bladder was distended as the result of obstruction of the common bile-duct by growth. *See Insp.* 1879, No. 54.

1398 Ulceration and Perforation of the Gall-Bladder.

A gall-bladder laid open to shew upon its anterior wall several small ulcers and one more considerable excavation, the latter being crossed by bands of persistent mucous membrane. At one part the floor of the excavation is formed by the adherent liver, at another the ulcerative process has established a communication between the bladder and the peritoneal cavity. There is recent lymph upon the serous coat. It was supposed that the ulceration was due to the presence of gall-stones. *Presented by Mr. E. Pye-Smith.*

1399 Cholecysto-Duodenal Fistula. Gall-Stone impacted in the Ileum.

A thickened and contracted gall-bladder with the duodenum, to which it is firmly adherent. Between the two viscera there is a fistulous communication easily admitting the middle finger. The opening in the intestine is situated about an inch from the pyloric ring. Below is mounted a portion of ileum in which is firmly impacted a gall-stone measuring an inch and a half in length by one inch in breadth.

Emma M., æt. 59, was admitted under Dr. Hale-White for intestinal obstruction. She had suffered from obstipation, vomiting, and abdominal pain for six days. Two days after admission laparotomy was performed and an artificial anus established in the small intestine. The patient died six hours later, and at the autopsy the stone was found to be impacted 33 inches above the ileo-cæcal valve. *See Insp.* 1895, No. 24.

1400   **Cholecysto-Duodenal Fistula.**

A gall-bladder, to the under surface of which a small portion of the duodenum is firmly adherent. A yellow rod indicates a fistulous communication between the bowel and the bladder. The walls of the latter are thickened, its cavity is contracted, and at its fundus are seen several small black concretions lodged in sacculi.

Presented by Dr. Stroud.

1401   **Cicatrix in the Gall-Bladder.**

A gall-bladder laid open to shew towards its neck a transverse puckered scar.

William B., æt. 50, was admitted under Dr. Babington in 1827 for progressive anæmia with emaciation, and died from chronic nephritis. At the autopsy several cysts were found in the pancreas, and the gall-bladder contained dark bile with black friable débris. *See Insp.* vol. 4, p. 92 ; and *Prep.* 1991 (2nd Edit.).

1402   **Calcified Gall-Bladder.**

A gall-bladder globular in shape, and measuring three and a half inches in diameter. It has been laid open to shew its interior encrusted with calcareous material. Its wall is thickened and its serous surface is roughened by old adhesions. A gall-stone is impacted at the orifice of the hypertrophied cystic duct. Another stone, which lies at the bottom of the preparation-bottle, was found in the gall-bladder, together with much inspissated pus.

From a middle-aged woman who was brought to the hospital dead. At the autopsy a dissecting aneurysm of the aorta was found.

1403   Adipose Gall-Bladder.

A gall-bladder laid open to shew its walls infiltrated with fat, which lies between the serous and submucous coats, and measures in parts one third of an inch in thickness.

Edward II., æt. 66, was admitted under Dr. Habershon in a moribund condition, and died the next day. At the autopsy the cellular tissue throughout the body was found to be loaded with fat, the kidneys were granular, and the liver cirrhotic. *See Insp.* 1872, No. 253.

1404   Papillomata of the Gall-Bladder.

A portion of a liver with the gall-bladder laid open to shew its mucous membrane thickly beset by papillomatous growths, which are small and sessile towards the fundus, whilst towards the neck they are larger and pedunculated.

Anne C., æt. 59, was admitted under Dr. Pavy, and died from phthisis. At the autopsy two large faceted calculi and some fragments of another were found in the gall-bladder. *See Insp.* 1881, No. 367 ; and *New Syden. Soc. Atlas of Path.* Fasc. 5.

1405   Warty Gall-Bladder.

The fundus of a gall-bladder everted to shew its mucous membrane covered by numerous slender and short papillary processes, which under the microscope are seen to consist of connective tissue covered by epithelium.

From Fanny L., æt. 52, who died from apoplexy.

1406   Cylindrical-celled Carcinoma of the Gall-Bladder.

A portion of the anterior edge of a liver, shewing in the situation of the gall-bladder a large mass of malignant growth, with which the omentum is firmly united. At the junction of the mass with the edge of the liver there is an opening made by operation into the cavity of the gall-bladder, which contains several calculi. On the reverse of the specimen is seen the upper part of the duodenum, the mucous membrane of which is stretched over the growth and ulcerated for a distance of three inches, beginning half an inch from the pylorus. The cut surface of the liver shews several secondary deposits, which histologically have the characters of cylindrical-celled carcinoma, some of the cells of which have undergone slight colloid change.

Fanny P., æt. 37, was admitted under Mr. Lane with an abdominal tumour in the right hypochondriac region, having previously suffered from occasional attacks of jaundice. Laparotomy was performed, and secondary deposits were observed in the liver. The gall-bladder, which was distended with bile, was incised, but no calculi were discovered. The patient died five weeks after the operation. Secondary deposits of growth were found in the peritoneum. *See Insp.* 1893, No. 411.

1407   Columnar-celled Carcinoma of the Gall-Bladder.

The anterior edge of the right lobe of a liver, with the gall-bladder and the hepatic flexure of the colon. The gall-bladder forms a globular mass five and a half inches in diameter, to the upper part of which the colon is adherent, a large ulcerated opening between the two cavities being indicated by a yellow rod. On the reverse of the specimen the enlargement of the gall-bladder is seen to be due to its infiltration by a soft white growth which has extended to the liver. The interior of the gall-bladder is covered with villous processes and irregular ragged sloughs. Histologically the growth is a columnar-celled carcinoma.

Thomas H., æt. 43, was admitted under Dr. Moxon with jaundice, vomiting, and a large tumour in the region of the gall-bladder. He had been ill for six months, and died, comatose, 25 days after admission. At the autopsy the liver was found to be greatly enlarged and to be almost entirely occupied by masses of secondary growth. *See Insp.* 1867, No. 35.

1408   Carcinoma of the Gall-Bladder.

A portion of the right lobe of a liver with the gall-bladder mounted to shew the latter surrounded by a mass of firm growth, to which is adherent a portion of the omentum and of the transverse colon. The primary mass of growth has invaded the adjacent substance of the liver, and there are secondary nodules scattered throughout the organ. Histologically the growth is a spheroidal-celled carcinoma.

Sarah C., æt. 65, was admitted under Dr. Perry with jaundice and a tumour in the right hypochondrium, having suffered from severe abdominal pain for the three months preceding her admission. She died three weeks later, and at the autopsy the gall-bladder was found to contain several gall-stones and there were secondary deposits of growth in the peritoneum. *See Insp.* 1892, No. 303.

1409   Carcinoma of the Gall-Bladder.

A portion of a liver with the gall-bladder, from the mucous membrane of which project several rounded masses of growth, the largest being about three quarters of an inch in diameter. Around the neck of the gall-bladder are several glands enlarged and infiltrated by malignant deposit which histologically has the characters of a spheroidal-celled carcinoma. The liver is granular upon the surface, and under the microscope shews marked cirrhosis.

George J., æt. 65, was admitted under Dr. Taylor with ascites and general dropsy, and died fourteen days after admission. At the autopsy the arteries were found to be extremely atheromatous, and the heart was hypertrophied and dilated. There was a carcinomatous ulcer in the descending colon, and there were numerous secondary deposits in the peritoneum and liver. *See Insp.* 1891, No. 178.

1410   Carcinoma of the Gall-Bladder.

A sagittal section of a liver shewing the gall-blader to be small, filled with faceted gall-stones, and surrounded by a mass of growth measuring four inches in diameter. The growth infiltrates the liver-substance, and in parts presents a ragged surface, the result of ulceration. Histologically it is a carcinoma, the alveoli of which are separated by a scanty stroma and filled with large spheroidal cells.

Julia C. æt. 46, was admitted under Dr. Habershon with a painful tumour on the right side of the abdomen, which had been noticed about four months. She became jaundiced and emaciated, and died two months after admission. At the autopsy a cavity containing several ounces of fœtid pus was found in the region of the gall-bladder, and the glands in the portal fissure were enlarged and infiltrated with secondary deposits. *See Insp.* 1867, No. 9.

1411   Carcinoma of the Gall-Bladder.

A gall-bladder laid open to shew upon its anterior wall a mass of growth projecting into its cavity, and resembling in size and shape a peeled walnut. There is an enlarged lymphatic gland by the side of the cystic duct. Histologically the growth is a spheroidal-celled carcinoma, with scanty stroma.

Presented by Dr. Stroud.

1412   Colloid Carcinoma of the Gall-Bladder and Bile-Ducts.

The gall-bladder with the structures in the portal fissure. The gall-bladder has been incised to shew its wall thickened by infiltration with soft gelatinous deposit. The main divisions of the bile-ducts and an enlarged gland at the neck of the bladder are similarly affected. Histologically the growth is spheroidal-celled carcinoma, most of the cells of which have undergone colloid change.

From Mary W., æt. 73. During life her gall-bladder was felt to be enlarged, and at the autopsy was found to contain a large calculus.

Presented by Mr. Ewen, 1857.

1413   Gall-Bladder filled with Gall-Stones.

A gall-bladder incised to shew the cavity tightly packed with numerous faceted gall-stones. A calculus about the size of a pea is lodged in the cystic duct a little beyond the neck of the gall-bladder.

From a patient aged 68.

Presented by Mr. T. E. Bryant.

1414   Gall-Bladder filled with Calculi.

A portion of a liver with the gall-bladder laid open to shew its cavity slightly enlarged and completely filled by a large number of brown faceted gall-stones.

1415   Calculus impacted in the Neck of the Gall-Bladder.

A gall-bladder laid open to shew its cavity somewhat dilated as the result of obstruction by a calculus, which is seen on the reverse of the specimen to be impacted in its neck. In the recent state the bladder contained three and a half ounces of thin colourless fluid.

Harriet V., æt. 37, was admitted under Dr. Rees with symptoms of stone in the kidney, from which she had suffered for many years. She died the day after admission, and at the autopsy both kidneys were found to be in a condition of calculous pyonephrosis. *See Insp.* 1868, No. 279.

1416   Gall-Bladder containing colourless Mucus.

A gall-bladder which in the recent state contained colourless fluid. There was no obstruction to the ducts.

James II., æt. 13, was admitted under Mr. Cock for acute necrosis of the tibia, and died from pyæmia two weeks after the onset of symptoms. At the autopsy the liver was found to weigh 50 ounces, and the contents of the alimentary canal were free from bile. *See Insp.* 1869, No. 94.

SECTION XXVII.—INJURIES AND DISEASES
OF THE BILE-DUCTS.

Laceration: 1417.
Dilatation: 1418-1421.
Impacted Calculi: 1422-1425.
Stricture: 1426.
Carcinoma: 1427, 1428.
Papilliferous Cyst: 1429.
Fistulæ: 1430.

1417   **Laceration of the Hepatic Duct.**

A gall-bladder and the main bile-ducts laid open to
shew at the origin of the hepatic duct a longitudinal
laceration in its wall, which communicates by a short
track through the connective tissue of the portal fissure
with the peritoneal cavity. The yellow rods are placed
in the ducts leading from the right and left lobes of the
organ, and the red rod passes through the sinus above
mentioned. There is considerable ecchymosis of the
adjacent liver.

> William B., æt. 40, was admitted under Mr. Bryant, having
> been struck in the abdomen by a stone 2 tons in weight which
> was being raised by a crane. A week later he died from perito-
> nitis, and at the autopsy 2 pints of bile-stained fluid containing
> blood-clots were found in the peritoneal cavity. There were
> several superficial lacerations of the liver. *See Insp.* 1887,
> No. 355.

1418   **Dilated Hepatic Duct. Ascending Suppurative
Hepatitis.**

An antero-posterior section through a liver at the
situation of the gall-bladder, with the adjacent portion
of the duodenum. The gall-bladder is seen on section

to be greatly contracted and its cystic duct, indicated
by a yellow rod, leads into a cavity two inches across
produced by the dilatation of the hepatic duct, which in
the recent state contained bile and pus together with a
great number of small black calculi. The common
bile-duct, indicated by a red rod, and also communicating with this cavity, is somewhat dilated and contained a gall-stone about the size of a cherry impacted
half an inch from the papilla. Scattered through the
substance of the liver and beneath its capsule are many
small ragged abscess-cavities.

Louisa W., æt. 30, was admitted under Dr. Pitt for jaundice
of 14 days' duration. Five months previously, after a severe
attack of enterica, a tumour had been noticed in the region of the
gall-bladder and had gradually disappeared. On admission the
liver was found to be uniformly enlarged, and the patient suffered
from pyrexia and rigors. She died 14 days later, and at the
autopsy the liver weighed 110 ounces and the peritoneal cavity
contained a considerable quantity of bile-stained purulent lymph.
See Insp. 1894, No. 458.

1419   Dilatation of the Common Bile-Duct.

A portion of a liver shewing upon its under surface a
thick-walled cyst measuring about six inches in diameter. The interior of the cyst is smooth, and
presents three openings, communicating respectively
with the dilated hepatic and cystic ducts and with
the distal portion of the common duct, indicated by
red, blue, and yellow rods. The last three-quarters
of an inch of the common duct is of less than the
normal calibre, and presents a valvular fold so far
obstructing its lumen that in the recent state fluid
could not be forced from the cyst through the biliary
papilla.

From Alice O., æt. 21, who suffered from persistent jaundice
for two and a half years, the onset of which was not preceded by
pain. A tumour in the hepatic region, extending to the level of
the umbilicus, was twice aspirated, three and a half pints of bile

being on each occasion withdrawn. Immediately after the second
operation the gall-bladder was laid open and stitched to the abdominal wall. Forty-eight hours later the patient died. At the
autopsy no gall-stone was discovered.

Presented by Dr. Farrant Fry, 1887.

1420   Dilated Bile-Ducts.

A portion of a liver, which has been incised to shew
the main branches of the bile-ducts traversing its substance to be greatly dilated, as the result of obstruction
by a small cancerous growth situated at the commencement of the ductus communis choledochus.

Robert B., æt. 68, was admitted under Dr. Habershon for
jaundice of three weeks' duration, and died three weeks later from
hæmatemesis and melæna. At the autopsy the dilated ducts
within the liver were found to be filled with a clear mucoid fluid
measuring about a pint. The gall-bladder contained a calculus.
See Insp. 1858, No. 123.

1421   Dilated Bile-Ducts.

A portion of a liver with the gall-bladder, main bile-
ducts, and duodenum. The common bile-duct is thickened and dilated, measuring an inch and a quarter in
circumference, and its lining-membrane in the recent
state presented numerous yellow patches regarded as
xanthelasmic. The gall-bladder is shrunken, and contained a small faceted calculus. Its duct is shortened
and there is a depressed scar on the anterior surface of
the portion of the liver overlying it. The biliary papilla
is patent, and immediately above it the interior of the
duct is scarred.

Louisa B., æt. 48, was admitted under Dr. Wilks for jaundice
and xanthelasma, the former symptom having persisted since an
attack of biliary colic twelve months previously. She died ten
days later from facial erysipelas, and at the autopsy the liver was
found to weigh forty-eight ounces, and the xanthelasmic change
was observed to extend into the smallest visible radicles of the
bile-ducts. The capsule of the spleen was covered by opaque
spots. *See Insp.* 1872, No. 246; and *Drawing*, 157(63).

1422   Biliary Obstruction. Chole-cysto-enterostomy.

A portion of a liver with the gall-bladder and the main bile-ducts, shewing the common bile-duct to be occupied by a black calculus measuring three-quarters of an inch in diameter, and reaching to within an inch of the biliary papilla. The calculus presents a bifurcation at its proximal end corresponding to the junction of the cystic with the main hepatic duct, both of which are thus partially obstructed. The gall-bladder and the cystic duct are dilated, as also are the hepatic ducts throughout the liver. The margins of an opening in the fundus of the bladder are firmly sutured to the border of a corresponding opening in the small intestine, in which is seen a perforation, indicated by a red rod, just beyond the line of union.

Samuel W., æt. 41, was admitted under Dr. Goodhart for enlargement of the liver and spleen with jaundice, from occasional attacks of which he had suffered for fourteen years. An exploratory operation was performed by Mr. Lane, and as it was found impossible to remove the gall-stones, an anastomosis was established between the fundus of the gall-bladder and the jejunum. Seven weeks after the operation the patient died, and at the autopsy general suppurative peritonitis was discovered, and the left pleural cavity contained some sero-purulent fluid. The spleen weighed 53 ounces. *See Insp.* 1894, No. 312.

1423   Gall-stones. Biliary Obstruction. Chole-cysto-colic Fistula.

A portion of a liver with the gall-bladder and main bile-ducts, the duodenum, and the hepatic flexure of the colon. The gall-bladder is adherent to the liver, thickened and contracted, and within it is seen a gall-stone. The colon is adherent to the fundus of the gall-bladder, and communicates with it by several ulcerated openings. The ductus communis choledochus is dilated so as to readily admit the middle finger, and at its termination in the duodenum is lodged a calculus. An inch and a half above the biliary papilla is seen an

F 2

opening in the wall of the duodenum leading into the dilated duct above the situation of the stone.

Thomas H., æt. 60, was admitted under Dr. Taylor for slight jaundice of 16 months' duration, with enlargement of the liver. Fifteen weeks later he died, and at the autopsy the body was found to be deeply jaundiced, and there was tuberculous disease of the meninges of the brain, of the lungs, pericardium, peritoneum, and spleen. *See Insp.* 1883, No. 198.

1424   **Calculus impacted in the Common Bile-Duct.**

The terminal portions of the cystic and hepatic ducts with the common bile-duct and the upper part of the duodenum. The common duct has been laid open to shew a black calculus as large as a cherry, firmly impacted just above the biliary papilla. Above the obstruction the ducts are greatly dilated.

Anne C., æt. 46, was admitted under Dr. Bright in 1837, and died from hæmoptysis due to chronic phthisis. At the autopsy the liver and spleen were found to be enlarged, and the kidneys were granular. *See Insp.* vol. 24, p. 24.

1425   **Calculus impacted at the Biliary Papilla.**

A portion of a duodenum, with the common bile-duct laid open to shew a calculus about the size of a large pea impacted in the duct immediately above the biliary papilla. The walls of the duct are thickened, and its lumen is dilated so as to admit the little finger.

1426   **Carcinomatous Stricture of the Bile-Ducts.**

A section of a liver through the portal fissure shewing the ducts within the organ to be greatly dilated. The commencement of the hepatic duct and its main branches are surrounded and constricted by a white fibrous mass about an inch long and half an inch in width. Histological examination shews that this adventitious material infiltrates the wall of the ducts, and consists of cylindrical-celled carcinoma, and that the liver itself is cirrhotic.

George Q., æt. 32, was admitted under Dr. Moxon for jaundice, xanthelasma, and enlargement of the liver, having during the preceding ten months suffered on several occasions from attacks of pain resembling biliary colic. Hæmatemesis and epistaxis supervened, and the patient died about a month after admission. At the autopsy the gall-bladder and bile-ducts were found to be filled with colourless watery mucus. See Insp. 1872, No. 169; and Trans. Path. Soc. vol. 24, p. 129.

1427   Carcinoma of the Common Bile-Duct.

The common bile-duct, with the duodenum and the head of the pancreas. The duct has been laid open to shew its wall for a distance of one inch from the biliary papilla infiltrated by firm white carcinomatous deposit, which narrows and obstructs its lumen. The growth has a well-defined margin, above which the duct is dilated and measures one and a quarter inches in circumference. Histologically the growth is composed of narrow branching alveoli containing very large epithelial cells.

Robert R., æt. 70, was admitted under Dr. Pavy with jaundice of four months' duration, and died in a state of coma two months after admission. The gall-bladder was felt to be enlarged, and at the autopsy was found to measure five inches in length and to be distended with dark-green bile. There were secondary deposits in the liver and kidney, and the hepatic tissue was cirrhotic. See Insp. 1889, No. 85.

1428   Carcinoma of the Bile-Ducts.

A portion of a liver with the gall-bladder laid open to shew a calculus of the size of a mulberry lodged at its neck. Around the calculus and in the transverse fissure is a mass of hard growth, which has invaded and obstructed the main hepatic ducts, and appears upon the anterior surface of the organ. Histologically the growth is a spheroidal-celled carcinoma.

Maria R., æt. 64, was admitted under Dr. Pavy for jaundice of six weeks' duration. The gall-bladder was felt to be enlarged, and a mass was discovered in the umbilical region. The jaundice became more intense, and the patient died seven months after admission. At the autopsy the gall-bladder was found to contain four ounces of colourless mucoid fluid with about 30 faceted gall-stones. The lower lobe of the right lung was consolidated by hypostatic pneumonia. *See Insp.* 1883, No. 189.

1429   Papilliferous Cyst of the Common Bile-Duct.

A liver, to the under surface of which is attached a large thick-walled cyst which has been laid open to shew its interior beset by numerous polypoid outgrowths. These cauliflower-like processes are seen under the microscope to consist of a richly nucleated connective tissue covered by ill-defined necrotic epithelium. On the reverse of the specimen the first part of the duodenum presents several perforations by which it communicates with the cavity of the cyst.

James W., æt. 4, was admitted under Dr. Moxon for swelling of the abdomen, emaciation, and vomiting of seven months' duration. The right half of the abdomen was occupied by a fluctuating tumour from which five pints of greenish purulent fluid were withdrawn. The patient died 11 days after the operation, and at the autopsy the cyst was found to communicate with the cystic, hepatic, and common bile-ducts, and with the fundus of the gall-bladder. *See Insp.* 1871, No. 261.

1430   Fistula between Common Bile-Duct and Duodenum.

A gall-bladder, with the cystic and common bile-ducts and a portion of the duodenum. Rods have been passed along the pancreatic and biliary ducts into the duodenum, in which one third of an inch above the papilla is seen a small ulcerated opening between the bile-duct and the bowel. The gall-bladder is shrunken and its cavity sacculated.

Edward W., æt. 43, was admitted under Dr. Wilks for ascites and carcinoma of the stomach, and died six days after admission. Three years previously he had suffered from an attack of jaundice. At the autopsy the ducts throughout the liver were found to be much dilated, as were also the hepatic and common bile-ducts. *See Insp.* 1876, No. 323.

Section XXVIII.—GALL-STONES.

Biliary Sand : 1431.
Biliary Calculi : 1432–1446.
Passed per rectum : 1447–1452.
Discharged through a Fistula : 1453.
Removed by Operation : 1454, 1455.
Causing Intestinal Obstruction : 1455–1459.

1431   Biliary Sand.

A quantity of sabulous material, consisting of black grains of inspissated bile-pigment.

> George R., æt. 40, was admitted in 1828 with a dislocated clavicle and fractured ribs, and died ten days later from pleuro-pneumonia. At the autopsy the mucous membrane of the gall-bladder was scarred, and its cavity contained the minute black grains above described. *See Insp.* vol. 5, p. 138.

1432   Biliary Calculi.

Thirteen small black gall-stones, consisting chiefly of bile-pigment. They have a roughened nodular surface somewhat resembling that of a mulberry.

> Thomas F., æt. 60, was admitted under Dr. Taylor for chronic bronchitis with cardiac dilatation, from which he died. At the autopsy the gall-bladder and bile-ducts were found to be healthy. *See Insp.* 1887, No. 39.

1433   Biliary Calculi.

A dried gall-bladder, in the cavity of which are seen about twenty calculi pale in colour, and of the shape and size of small peas. They are composed chiefly of cholesterine.   *Presented by Mr. H. R. Hillier.*

1434   Biliary Calculus.

A pale yellow smooth gall-stone, oval in shape, and measuring an inch and a quarter in its longest diameter. It consists almost entirely of cholesterine.

Presented by Dr. Fagge, 1861.

1435   Biliary Calculus.

An ovoid calculus measuring an inch in length and half an inch in width, yellowish white in colour, and consisting chiefly of cholesterine. It has a mammillated mulberry-like surface.

1436   Biliary Calculus.

An oval calculus, one inch and three-quarters in length, which has been broken across to shew a yellow central nucleus consisting of plates of cholesterine arranged radially, surrounded by a white amorphous outer zone a quarter of an inch in thickness.

1437   Biliary Calculus.

An oval gall-stone about an inch in length, the cut surface of which shews a yellow nucleus having a radiating crystalline structure composed almost entirely of cholesterine. The crust is amorphous, and is brownish in colour from admixture with bile-pigment.

1438   Biliary Calculus.

A gall-stone, moulded to the shape of the gall-bladder, and presenting at one end a projection corresponding to the orifice of the cystic duct. The calculus is two and three-quarter inches in length, and was found in the body of an elderly lady.

Presented by Mr. Callaway.

1439   Biliary Calculi.

A dried gall-bladder containing three large calculi. Behind is mounted a numerous collection of faceted gall-stones, varying from a line to half an inch in diameter.   *Presented by Mr. T. E. Bryant.*

1440   Biliary Calculi.

A dried gall-bladder, containing about a hundred very small polygonal calculi.

Presented by Mr. H. R. Hillier.

1441   Biliary Calculi.

Several faceted gall-stones, some of which have been broken to display a brown nucleus of inspissated bile surrounded by a white zone of cholesterine and a thin slate-coloured crust containing, in addition, phosphate of lime and fat. The unbroken stones have roughly the shape of triangular prisms. They were analysed by Dr. Golding-Bird.

1442   Biliary Calculi.

A large number of faceted gall-stones, varying in size from a small pea to a hazel-nut. They were analysed by Dr. Golding-Bird and were found to be composed of cholesterine covered by biliary matter. The nucleus consisted of inspissated bile ; the external layer contained also phosphate of lime and fat.

Presented by Dr. Stroud, 1831.

1443   Biliary Calculi.

Numerous slate-coloured gall-stones, varying in size from a line to three-eighths of an inch in diameter, of irregular shape, some being faceted and others mammillated.

From the gall-bladder of a man, æt. 30, who died in 1828.

1444   Biliary Calculus.

A small oval gall-stone, brown in colour, and covered by a deposit of minute shining crystals.

Mary H., æt. 45, was admitted under Dr. Back in 1827 with jaundice, which before her death assumed a deep greenish tinge. She died seven weeks after admission, and at the autopsy carcinomatous nodules were found in the liver, the glands in the portal fissure were enlarged, and the hepatic ducts were dilated. *See Insp.* vol. 4, p. 124.

1445   Biliary Calculi.

Two black gall-stones, each about the size of a pea. They are very irregular on the surface and are apparently formed by the agglutination of smaller polygonal concretions of biliary pigment.

1446   Biliary Calculi from the Hepatic Ducts.

Several gall-stones, pale in colour and consisting chiefly of cholesterine, which were found in the hepatic ducts of a patient who died from carcinoma of the liver. *See Prep.* 1347.   *Presented by Mr. Callaway*, 1829.

1447   Biliary Calculus passed per Rectum.

An egg-shaped calculus, an inch and a half in length and an inch transversely at its widest part. It has been divided to shew beneath its brown exterior a uniform white substance consisting chiefly of cholesterine.

From a patient who passed the calculus after suffering for " many days from severe symptoms of local obstruction."

Presented by Mr. W. H. Turner.

1448   Gall-stone passed per Rectum.

A rounded gall-stone, measuring rather more than an inch in its longest diameter, which was passed by a patient who had previously suffered from symptoms of intestinal obstruction.   *Presented by Mr. Hearnden.*

1449   Gall-stone passed per Rectum.

A globular calculus presenting two flattened facets, one at either pole. It measures an inch and a quarter in diameter and weighs 270 grains. It consists chiefly of cholesterine, and is coated with a friable crust of biliary pigment.

From a lady, æt. 78, by whom it was evacuated per rectum after seven days' constipation. See *Trans. Path. Soc.* vol. 38, p. 160.

Presented by Dr. Pye-Smith.

1450   Biliary Calculi passed per Rectum.

Two biliary calculi, one globular, measuring an inch in diameter, the other conical, measuring two inches from apex to base. The base of the cone is concave, and appears to have been moulded to the surface of the sphere. Each calculus has been bisected to shew it to be composed of concentric layers of cholesterine.

From a middle-aged lady, by whom they were passed per anum.

Presented by Mr. T. Newington.

1451   Biliary Calculus passed per Rectum.

The two halves of a cylindrical calculus with rounded ends measuring nearly three inches in length and an inch and a quarter in diameter. The fractured surface shews a core of brown material surrounded by a zone one-eighth of an inch in thickness composed of radiating cholesterine crystals. Its outer surface is brown, and in parts mammillated.

From a lady, æt. 60, who passed the fragments above-described after complete obstruction of ten days' duration associated with little vomiting and moderate pain. See *Trans. Path. Soc.* vol. 38, p. 162.

Presented by Dr. Pye-Smith.

1452   **Biliary Calculi.**

Seven white round gall-stones varying in size from a
sixteenth to an eighth of an inch in diameter. They
were found in the fæces.

Presented by Mr. T. E. Bryant.

1453   **Biliary Calculi discharged externally.**

Two faceted gall-stones, each about the size of a cherry,
which were discharged through a sinus opening at the
umbilicus.

From a female patient.

Presented by Mr. Callaway.

1454   **Biliary Calculus removed by operation.**

An oval gall-stone, an inch and a quarter long and
three-quarters of an inch broad, which was removed
through a fistulous opening.

Presented by Mr. Bryant, 1878.

1455   **Gall-stones removed from the Ileum by opera-
        tion.**

A. small portion of ileum, presenting upon its free
border an incision two inches in length, the edges of
which are united by a continuous catgut suture.
Through this incision was removed the gall-stone, which
rests on the bottom of the preparation-jar. The calculus
is nearly two inches long, rather more than one inch in
diameter at its broadest part, weighs 238 grains, and is
moulded to the shape of the gall-bladder. Above is
mounted a portion of the liver, the gall-bladder, and
the first part of the duodenum. The gall-bladder is
thickened and contracted, and there is a fistulous communication between it and the bowel, the parts being
united by firm adhesions. The anterior edge of the liver
is thin and bent back upon the upper surface of the
organ.

Eliza R., æt. 50, was admitted under Mr. Bryant with symptoms of acute intestinal obstruction of three days' duration, never having previously suffered from any illness except occasional dyspepsia. Laparotomy was performed, and the peritoneum found to be acutely inflamed. The patient died 70 hours after the operation, and at the autopsy the incision in the ileum was found to be 12 inches above the cæcum. *See Insp.* 1878, No. 301 ; and *Trans. Clin. Soc.* 1879, p. 106.

1456 Biliary Calculi.

Three large yellow calculi, each weighing about 55 grains. They are cubical in shape and have a smooth surface.

From Mrs. D., who died with symptoms of intestinal obstruction of six days' duration, associated with a tumour in the right hypochondrium. At the autopsy the right lobe of the liver was found to be occupied by a carcinomatous mass, beneath which the distended gall-bladder was pressed firmly against the pylorus and first part of the duodenum. The bowel in this situation was black and gangrenous. The cystic duct was obliterated, and in addition to the gall-stones which form the preparation, numerous smaller ones were found in the bladder. *See Note Book*, 2, p. 37.

Presented by Dr Tyerman, 1807.

1457 Gall-stone impacted in the Ileum.

A piece of ileum, the wall of which has been partly removed to shew impacted in it an oval calculus measuring an inch and three-quarters in length. Its impaction caused fatal intestinal obstruction.

1458 Biliary Calculus producing Intestinal Obstruction.

An oval gall-stone, one inch and three-quarters in length and a little more than an inch in its transverse diameter. Its surface is brown and scaly.

From Mrs. S., a very stout woman, æt. 59, who three months before her fatal illness suffered from pain in the side and fever. Six days before death vomiting came on, but there was neither abdominal pain nor tympanites. At the autopsy the gall-stone was found firmly impacted about thirty inches below the pylorus. There was a communication below the biliary papilla between the duodenum and the gall-bladder, which were firmly adherent to each other.

Presented by Mr. Pye-Smith.

1459   **Biliary Calculi causing Intestinal Obstruction.**

Two gall-stones, one of them cylindrical with faceted ends, the other, which is rather smaller, polygonal in shape. The larger measures one inch in diameter. They are reported to have caused fatal obstruction.

Presented by Dr. Addison.

Section XXIX.—INJURIES AND DISEASES OF THE PANCREAS.

1460   Laceration of the Pancreas.

A pancreas mounted to shew a complete transverse laceration of its body corresponding to the situation where the gland crossed the vertebral column.

> Charles W., æt. 25, was admitted under Mr. Cock having been run over by a cart. He died five days after the accident, and at the autopsy the stomach, spleen, and left kidney were found to be lacerated. There was peritonitis limited to the lesser sac of the peritoneum. *See Insp.* 1867, No. 287.

1461   Accessory Pancreas.

A portion of a duodenum with the head of the pancreas, and an accessory pancreas. The normal pancreatic duct is indicated by a pink rod appearing at the orifice of the biliary papilla, and a bristle has been passed through a second duct leading from the accessory pancreas into the duodenum at a point an inch and a half below the papilla.

> James T., æt. 34, was admitted in 1832 and died from Asiatic cholera. *See Insp.* vol. 16, p. 132.

1462   Pancreatic Calculi.

Two small oval calculi, white and granular on the surface. They were analysed by Dr. Golding-Bird and found to consist of oxalate of lime, phosphate of lime, and phosphate of magnesium.

Presented by Mr. Hilton.

1463   Pancreatic Calculus.

A small white calculus from the pancreatic duct.

1464   Impacted Calculi.   Dilated Pancreatic Duct.

The head of a pancreas and the adjacent portion of the duodenum. The pancreatic duct has been laid open and is seen to be dilated so as to measure nearly an inch in circumference. In its proximal portion there are lodged several white nodular calculi, one of which on the reverse of the specimen is seen to block the orifice of the duct at its point of junction with the common bile-duct, the latter being somewhat dilated.

James A., æt. 48, was admitted under Dr. Babington in 1842 for diabetes, and died in an epileptic fit 5 days after his admission. At the autopsy the pancreas was found to be small and indurated. *See Insp.* vol. 31, p. 230.

1465   Abscess in the Head of the Pancreas.

A pancreas with the gall-bladder and a portion of the duodenum. The gland has been incised to shew at its head and immediately in contact with the common bile-duct a cavity which in the recent state " would contain a small plum and had a shred of slough hanging in it." Close to the biliary papilla is seen the opening by which the abscess discharged its contents into the duodenum.

Emma K., æt. 28, was admitted under Dr. Habershon for ascites anasarca and albuminuria. About a month later jaundice came on and gradually disappeared in the course of the next fortnight. She died 3 months after admission, and at the autopsy the kidneys and spleen were lardaceous and the left kidney contained a large gumma. It was thought that the bursting of the abscess, which was regarded as a softening syphiloma, in the head of the pancreas had relieved the common bile-duct from pressure and led to the disappearance of jaundice. *See Insp.* 1867, No. 234; and *Guy's Hosp. Reps.* 1867, p. 391.

1466   Sarcoma of the Pancreas.

A portion of a pancreas from the surface of which projects a nodule of growth about the size of a filbert. It is partially embedded in the substance of the organ and histologically has the characters of a small round-celled sarcoma.

Ellen B., æt. 2, was admitted under Mr. Birkett for pain and swelling in the right shoulder, attributed to an injury received 7 weeks previously. The tumour rapidly increased in size and the patient died 3 weeks after admission. At the autopsy malignant growth was found in the humerus, scapula, skull, ribs, and lungs. *See Insp.* 1855, No 104; and *Prep.* 1098 (10) [2nd Edit.].

1467   Sarcoma of the Head of the Pancreas.

A pancreas with the common bile-duct and a portion of the duodenum. The head of the pancreas is occupied by a firm globular tumour having a nodulated surface and measuring three inches in diameter. The distal part of the pancreatic duct is much dilated and its proximal end is lost in the substance of the growth, which also surrounds but does not apparently compress the common bile-duct. The duodenum is unaffected. Histologically the growth is a vascular sarcoma consisting of round and spindle cells.

1468   Sarcomatous Glands around the Pancreas.

A mass consisting of pancreatic tissue infiltrated with malignant deposit and surrounded by enlarged lymphatic glands which are similarly affected. On histological examination the growth is found to be a sarcoma with oval and spindle cells.

Henry P., æt. 18, was admitted under Mr. Hilton with an ulcerating mass of malignant growth in the neck. Deposits were also detected in the skin and in the abdominal cavity. He died five weeks after admission, and at the autopsy growth was found in the cranium and dura mater, the ribs, heart, liver, kidneys, and spleen. *See Insp.* 1862, No. 153; and *Prep.* 1399 (65) [2nd Edit.].

1469   **Carcinoma of the Pancreas. Dilated Duct.**

The tail and a portion of the body of a pancreas, the latter infiltrated with carcinomatous deposit and the former occupied by a smooth-walled cyst measuring an inch and a half in diameter, which appears to be formed by dilatation of the pancreatic duct. Histologically the growth has the structure of cylindrical-celled carcinoma with much fibrous stroma.

Charles S., æt. 55, was admitted under Dr. Addison for wasting and abdominal distension, his illness having commenced six months previously. He died a month after admission, and at the autopsy the abdominal cavity contained a few pints of dark-coloured serum and the surface of the peritoneum was studded with cancerous nodules. It was thought that the disease had its origin in the body of the pancreas. There were secondary deposits in the mesenteric glands, the liver, and the pleura. *See Insp.* 1859, No. 51.

1470   **Carcinoma of the Pancreas. Dilated Duct.**

A pancreas the head of which with the surrounding lymphatic glands is occupied by a mass of growth having the histological characters of a carcinoma. The growth has invaded and obstructed the duct for about an inch from its orifice, the portion behind the obstruction being dilated.

Presented by Dr. Bright.

1471   **A Cyst in the Pancreas.**

A portion of a pancreas in the head of which is seen a cyst about three-quarters of an inch in diameter, the interior of which is smooth and fibrous. In the recent state the ducts of the gland were found to contain several branching calculi, one of which is at the bottom of the preparation-glass.

Daniel L., æt. 55, was admitted under Dr. Goodhart for bronchitis and cardiac dilatation, from which ten days later he died. *See Insp.* 1892, No. 28.

1472   Cysts in the Pancreas.

A pancreas presenting two cysts, one at the head the other towards the tail of the gland. The former contains a mass of blood-clot about as large as a hazel-nut, the latter in the recent state was filled with a fluid resembling turbid saliva. The walls of the cyst are smooth and fibrous.

William B., æt. 50, was admitted under Dr. Babington in 1827 for anæmia and general weakness, having eighteen months previously suffered from apoplexy and hemiplegia. At the autopsy the kidneys were found to be granular and the urine was coagulable by heat. The gall-bladder presented a cicatrix on its mucous membrane. *See Insp.* vol. 4, p. 92 ; and *Prep.* 1401.

1473   Cystic Disease of the Pancreas.

A pancreas shewing several small cysts with thin fibrous walls. In the recent state some 8 or 9 such cysts were observed varying in size from a hempseed to a hazel-nut and containing clear fluid.

Alexander D., æt. 27, was admitted under Dr. Mahomed with symptoms of cerebral tumour of two months' duration. He died 40 days after admission, and at the autopsy a cyst was found in the cerebellum, and there were numerous cysts in the kidney. *See Insp.* 1884, No. 127 ; and *Prep.* 1576 (74 A) [2nd Edit.].

1474   Pancreatic Cyst.

A stomach with the duodenum, transverse colon, and a portion of the anterior abdominal wall. An opening indicated by a yellow rod has been made through the parietes and leads by a channel between the colon and

G 2

the greater curvature of the stomach into a large cavity the relations of which are displayed on the lateral aspect of the preparation. The cyst occupies the position of the pancreas and has pushed forward the stomach and small omentum. Its walls are thick and fibrous and its inner surface is rough and in parts encrusted with calcareous material. On the reverse of the specimen the duodenum is seen stretched over the back of the sac. Histological examination of the dense wall of the cyst shews it to contain some minute fragments of pancreatic tissue.

John M., æt. 68, was admitted under Dr. Goodhart with a tumour situated between the ensiform cartilage and the umbilicus, and measuring 4 or 5 inches transversely. It was resonant on percussion, with non-expansile pulsation, and painless. On the day after admission Mr. Lane evacuated by incision 6 pints of thick yellowish-red fluid in which crystals of leucine and tyrosine were detected, together with a ferment. Eight days later the patient died from bronchitis and cardiac failure. At intervals during his illness the urine contained sugar. *See Insp.* 1894, No. 412.

1475   **Hydatid at the Head of the Pancreas.**

A portion of a liver mounted to shew in the situation of the head of the pancreas a globular cyst which in the recent state measured $3\frac{1}{2}$ inches in diameter, and contained 250 c.c. of clear hydatid fluid together with the membrane seen at the bottom of the preparation-glass. The duodenum is displaced and stretched over the cyst, and the common bile-duct and portal vein, indicated by yellow and blue rods respectively, are seen running in its wall. The gall-bladder is distended with bile and the surface of the liver is granular, the hepatic structure shewing under the microscope a considerable increase of fibrous tissue around the lobules and between the cells.

Percy A., æt. 6, was admitted under Dr. Taylor for persistent jaundice of 8 months' duration. On admission the liver was found to be enlarged and there was a globular tumour in the region of the gall-bladder. Ascites supervened for the relief of which paracentesis was performed. The patient died seven weeks after admission, and at the autopsy it was found to be impossible to force the bile from the gall-bladder into the duodenum until the cyst was opened. *See Insp.* 1894, No. 489.

1476   **Fibrous Stricture of the Biliary Papilla. Dilated Pancreatic Duct.**

A pancreas with the common bile-duct and a portion of the duodenum. In the duodenum at the situation of the biliary papilla is a small area denuded of mucous membrane, which in the recent state formed part of the wall of a submucous abscess. The biliary papilla is surrounded and narrowed by fibrous tissue and both the common bile-duct and the pancreatic duct are dilated, the former, indicated by a yellow rod, measuring one inch and a half in circumference. The pancreatic duct, indicated by a blue rod, presents a number of saccular and fusiform dilatations partially separated from each other by thin fibrous septa.

Frederick E., æt. 50, was admitted under Mr. Symonds with jaundice, for the relief of which cholecystotomy was performed. No calculi were discovered, and a biliary fistula was established. Two months later a second operation took place to effect a communication between the gall-bladder and the small intestine. Subsequently the original wound broke down and a fæcal fistula resulted. The patient died 3 weeks after the second operation, and at the autopsy one large and several smaller abscesses were found in the liver and there were localised collections of pus between adherent coils of intestine. *See Insp.* 1888, No. 200.

1477   **Pancreatic Duct dilated from Carcinomatous Obstruction.**

A pancreas which has been laid open to shew considerable dilatation and sacculation of its duct.

Patrick —, was admitted under Dr. Habershon for jaundice of ten weeks' duration, with enlargement of the liver. He died 9 weeks after admission, and at the autopsy a small nodule of scirrhous spheroidal-celled carcinoma was found in the head of the pancreas which completely obstructed Wirsung's canal and caused s me narrowing at the junction of the cystic and hepatic ducts. There was tuberculous excavation at the apices of the lungs, and the left kidney was in a condition of suppuration. *See Insp.* 1866, No. 286.

1478   Pancreatic Duct opened by a Gastric Ulcer.

The pyloric end of a stomach shewing a portion of a chronic ulcer in the base of which is a small orifice communicating with the main trunk of the pancreatic duct.

Presented by Dr. Bright.

Section XXX.—INJURIES AND DISEASES OF THE SPLEEN.

1479   Laceration of the Spleen.

A half of a spleen shewing upon its capsule a deep irregular laceration with shreds of lymph between its edges. The cut surface of the organ shews at the situation of the injury an effusion of blood beneath the capsule and a bruised condition of the splenic tissue.

Anne F., æt. 9, was admitted under Mr. Bransby Cooper in 1826, two days before her death, having been run over by a cart. The right thigh was fractured and the urine contained blood. At the autopsy the pelvis was fractured and the peritoneal cavity contained dark fluid blood. The bladder was ruptured. *See Insp.* vol. 1, p. 72.

1480   Laceration of the Spleen in process of Repair.

A portion of a spleen shewing an oblique rent almost dividing the organ. The fissure thus produced is filled with firm coagulum.

John M., æt. 34, was admitted under Mr. Hilton, having been jammed between a wheel of a cart and a wall. Three days later he died, and at the autopsy the skull, left arm, and several ribs were fractured. There was a large quantity of blood in the peritoneal cavity. *See Insp.* 1858, No. 210.

1481   Lacerations of the Spleen.

The half of a spleen mounted to shew several superficial lacerations of the capsule and of the subjacent tissue. The edges of the rents are united by organised thrombus.

Henry B., æt. 4, was admitted under Mr. Durham after a violent blow on the left side. On admission several ribs were found to be broken, and there was surgical emphysema. Six days later he died, and at the autopsy the pleural and peritoneal cavities were found to contain a considerable quantity of blood. The left lung was lacerated. *See Insp.* 1890, No. 266 ; and *Prep.* 215.

1482   Ruptured Spleen.   Hæmorrhage distending the Capsule.

A spleen with portions of the stomach and diaphragm, to which it is united by firm adhesions. The capsule has been partially removed and discloses a cavity which in the recent state contained blood-clot, portions of which are still seen adhering to the lacerated spleen-pulp.

Thomas S. was admitted under Mr. Morgan in 1830, having 12 days before his death been run over by a carriage. At the autopsy the second, third, and fourth ribs on the left side were found to be fractured and the peritoneal cavity contained a considerable quantity of bloody serum. There was a chronic ulcer in the lesser curvature of the stomach. *See Insp.* vol. 10, p. 36; and *Drawing,* 353.

1483 Spleen in Hydrochloric Acid Poisoning.

A portion of an enlarged spleen which has been uniformly blackened by the action of hydrochloric acid. In the recent state the cortical portion was hard and resistant, the central part being considerably softer.

Robert T., æt. 45, was admitted under Dr. Pye-Smith, having swallowed a considerable amount of hydrochloric acid with suicidal intent. He died an hour after taking the poison, and at the autopsy the alimentary tract as low as the middle of the jejunum was found to be corroded, and the fluid in the peritoneal cavity was strongly acid. *See Insp.* 1894, No. 454.

1484 Accessory Spleens.

Two small accessory spleens, one as large as a hazelnut and the other of the size of a pea, with a portion of the gastrosplenic omentum to which they are attached.

1485 An Accessory Spleen, enlarged, fibrous, and lardaceous.

An ovoid tumour measuring 6 inches in length and 3½ inches in thickness. It is enclosed in a fibrous capsule, and its cut surface exhibits interlacing strands of fibroid tissue together with sago-like grains of translucent material. Histologically the mass is seen to consist of a fibrous stroma with lardaceous malpighian corpuscles. Behind is displayed the relation of the tumour to the left kidney and the suprarenal body.

Enoch F., æt. 41, was admitted under Dr. Moxon for diarrhœa, having suffered from dysentery, contracted abroad, six years previously. His urine contained albumen, and a tumour was felt on the left hypochondriac region. A week after admission he died, and at the autopsy the mucous membrane of the colon was found to be blackened and scarred, and there was lardaceous

change in the intestine and kidneys. The spleen, which was also
lardaceous, weighed 3½ ounces. *See Insp.* 1874, No. 424; and
Prep. 2005 (80) [2nd Edit.].

1486   Spleen with Fissured Margins.

A small spleen weighing a little more than an ounce
and a half, and presenting upon its margins numerous
deep and irregular fissures.

Mary M., æt. 50, was admitted under Dr. Cholmeley in 1829
for chronic intestinal obstruction, from which she died. *See Insp.*
vol. 7, p. 137; *Preps.* 2259 (20) and 2022 (24) [2nd Edit.]; and
Drawings, 395 & 393.

1487   Foramen in the Spleen.

A portion of a spleen shewing a foramen lined by the
capsule of the organ and large enough to admit the
middle finger. Through this hole a piece of omentum
has passed and has become partially adherent.

1488   Atrophied Spleen.

A small spleen weighing rather less than an ounce, the
capsule of which is thickened and fibrous.

Richard C., æt. 41, was admitted under Dr. Mahomed for
chronic phthisis, from which he died. *See Insp.* 1883, No. 338.

1489   Enlarged Spleen.

A uniformly-enlarged spleen which histologically appears to be of normal structure.

Arabella T., æt. 15 months, was admitted under Dr. Hilton
Fagge for changes in the skeleton attributed to syphilis and
rickets. The spleen was enlarged and extended to the umbilicus.
The child died in convulsions. *See Insp.* 1882, No. 204; *Trans.
Path. Soc.* vol. 34, p. 197; and *Preps.* 1000 (23) to 1000 (27)
[2nd Edit.].

1490   Enlarged Spleen.

A uniformly enlarged spleen which measures 13 inches
in length, 8 inches across, and 4 inches in greatest
thickness, and weighs 81 ounces. Its capsule, especially

upon the convex surface, is marked by fibrous adhesions. Histologically the splenic tissue is apparently normal, and the blood-clot in the vessels contains a great excess of white corpuscles. *Presented by Dr. Curry.*

1491   Enlarged Spleen.

An enlarged spleen, the capsule of which is roughened by organised adhesions. The organ measures 12 inches long by 7 transversely, and in the recent state weighed 5 lbs. 14 ozs. Histologically the splenic tissue appears normal.

From Eliza T., æt. 41, who died in 1808. She is stated to have suffered from ascites and hypertrophy of the heart. *See Old Museum Book*, No. 100.

1492   Small Lardaceous Spleen.

A small spleen, firm and waxy, which on histological examination proves to be highly lardaceous.

1493   Lardaceous Spleen.

A portion of a spleen upon the cut surface of which are seen numerous translucent nodules somewhat resembling boiled sago-grains. Histologically they are found to consist of lardaceous malpighian corpuscles.

Anne O., æt. 47, was admitted under Mr. Cock for necrosis of the bones of the face due to syphilis contracted twenty years previously. She died 18 weeks after admission, and at the autopsy the kidneys were found to be in a condition of tubal nephritis, and the liver was scarred, fatty, and lardaceous. *See Insp.* 1855, No. 54.

1494   Lardaceous Spleen.

A section of a spleen in which the malpighian corpuscles are unusually evident, having been stained mahogany-brown by the action of iodine. Histological examination shews the presence of lardaceous deposit widely diffused throughout the organ.

James D., æt. 18, was admitted under Mr. Hilton with disease of the hip-joint, from which he had suffered four years. For three years before his death there had been a persistent discharge of pus from sinuses about the joint. At the autopsy the liver and kidneys were found to be lardaceous, and there was tuberculous pleurisy. *See Insp.* 1856, No. 71.

1495   Lardaceous Spleen.

A spleen, the cut surface of which has a mottled translucent appearance from the presence of lardaceous change chiefly affecting the malpighian bodies.

For history and reference see *Prep.* 1485.

1496   Capsulitis of the Spleen.

A spleen, the capsule of which is much thickened, partially separated from the parenchyma of the organ, and marked by numerous punched-out depressions. Between the depressions the surface of the capsule is smooth, glistening and free from adhesions.

1497   Capsulitis of the Spleen.

A section of a spleen, on the convex surface of which is seen an irregular pitted plaque of fibrous tissue. Elsewhere the capsule is roughened by adhesions.

1498   Local Capsulitis of the Spleen.

A portion of a spleen shewing upon its capsule a considerable depression with puckered edges. In the base of the depression is seen a white fibrous plaque of cartilaginous consistency about a line in thickness. Histologically the plaque consists of dense fibrous tissue, the subjacent parenchyma being engorged with blood.

Sarah K., æt. 30, was admitted under Mr. Bransby Cooper in 1827 for a psoas abscess, with chronic phthisis, from which about 8 months later she died. The left suprarenal capsule was considerably enlarged by caseous deposit. *See Insp.* vol. 4, p. 102.

1499 Local Capsulitis of the Spleen.

A portion of a spleen, on the surface of which is seen a smooth, slightly-raised fibrous patch with well-defined sinuous margins.

1500 Local Capsulitis of the Spleen.

A spleen, somewhat enlarged and shewing upon its convex surface a smooth cartilaginous plate about as large as the palm of the hand. Histologically it is found to have the structure of fibrous tissue.

1501 Gummatous Capsulitis of the Spleen.

A vertical section through a left lung, the diaphragm, and the spleen, shewing the latter organ to be considerably enlarged and to be surrounded by a dense layer of fibrous tissue, in parts as much as an inch in thickness. The fibrous material has infiltrated and destroyed the muscular tissue of the diaphragm, which is firmly adherent to the base of the lung. Histological examination of the inflammatory tissue shews it to be gummatous.

Thomas II., æt. 28, was admitted under Mr. Cooper Forster with balanitis and phimosis for which circumcision was performed. Pyæmia supervened, and the patient died 16 days after the operation. At the autopsy the testes were found to be gummatous, and there were scars upon the legs. *See Insp.* 1860, No. 147.

1502 Enlarged Spleen in Heart Disease.

A spleen, which in the recent state was hard and weighed 9 ounces. It is uniformly enlarged and its capsule presents a few short filamentous adhesions.

Emily T., æt. 30, was admitted under Dr. Goodhart for ascites and œdema with the physical signs of mitral regurgitation. Seven weeks later she died, and at the autopsy the heart was found to weigh 18½ ounces, and the mitral and tricuspid valves were fibrous, their orifices being somewhat constricted. *See Insp.* 1893, No. 439.

1503   Spleen in Ulcerative Endocarditis.

An enlarged spleen, weighing twenty-eight ounces, incised to shew at its left margin two small recent infarcts. Extending from the opposite margin, almost across the anterior surface, is seen a large triangular area roughened by adhesions and corresponding to the situation of a considerable infarct of older date.

Harry B., æt. 24, was admitted under Dr. Goodhart for heart disease and pleurisy. He had had frequent attacks of rheumatic fever, and while in the hospital he suffered from symptoms of embolism of the left brachial and left middle cerebral arteries. He died 3 months after admission, and at the autopsy the aortic and mitral valves were found to be affected with ulcerative endocarditis. *See Insp.* 1891, No. 320.

1504   Splenic Infarct.

A spleen, on the convex surface of which is seen a small, slightly depressed area, having a smooth surface clearly contrasting with the wrinkled capsule of the organ. The section shews that this area forms the base of a wedge-shaped infarct which is partially discoloured and surrounded by a hyperæmic zone.

James S., æt. 35, was admitted under Dr. Cholmeley in 1827 for dyspnœa, with rapid and irregular pulse and rigors. Three days later he died, and at the autopsy the mitral valve was found to be much contracted, and presented a ring of minute vegetations with "some little appearance of ulceration." *See Insp.* vol. 2, p. 32.

1505   Splenic Infarct.

A spleen, beneath the convex surface of which at its upper end is seen an infarct triangular on section, the base measuring three-quarters of an inch. The infarct is clearly defined from the surrounding spleen-tissue, and is of pale colour and solid consistency.

William H., æt. 46, was admitted under Dr. Cholmeley in 1827 for general anasarca and albuminuria. Whilst in the hospital he had repeated convulsions and died comatose a fortnight after his admission. At the autopsy the heart was found to be normal and the kidneys were in an "advanced stage of degeneration." *See Insp.* vol. 3, p. 64.

1506   Spleen scarred from Infarction.

A portion of a spleen, mounted to shew about the centre of its convex surface a depressed fibrous scar apparently resulting from previous infarction.

George L., æt. 38, was admitted under Dr. Addison in 1832, and died from pneumonia. At the autopsy the heart was found to be dilated and hypertrophied, and there were soft vegetations upon the mitral valve and upon the endocardium of the left auricle. The spleen was more than double its natural size, and one of the kidneys contained an infarct. *See Insp.* vol. 11, p. 172.

1507   Thrombosis of the Splenic Artery.

The hilum of a spleen, shewing the distal portion of the splenic artery to be tortuous, atheromatous, and occluded by thrombus. A small branch to the lower end of the organ, indicated by a red rod, is patent, and on the reverse of the specimen the part supplied by this twig is seen to have a normal spongy structure, the rest of the viscus being paler and of closer texture.

Ann R., æt. 36, was admitted under Dr. Moxon for hemiplegia, the onset of which had occurred suddenly 9 months before admission. She became aphasic, had severe headache, and passed into a state of coma in which one month later she died. Shortly before her death a painful tumour appeared on the left hypochondrium. At the autopsy the left middle cerebral artery was found to be occluded by thrombus, and there were patches of softening on the left side of the brain. The heart was healthy. The left internal jugular vein was plugged, and the neighbouring cervical glands were enlarged. *See Insp.* 1875, No. 272.

1508   Embolic Abscess of the Spleen.

A spleen divided longitudinally to shew beneath the capsule of its convex surface a cavity triangular in shape, which in the recent state contained greenish tenacious pus. At the upper part of the cavity is seen a mass about the size of a hazel-nut, consisting of necrotic and partially detached splenic tissue.

Alfred F., æt. 25, was admitted under Dr. Habershon with signs of heart disease and albuminuria. Seventeen years previously he had suffered from acute rheumatism. He died about 2 months after admission, having become hemiplegic and aphasic a week before his death. At the autopsy abscesses were found in the brain and liver and the kidneys shewed minute hæmorrhages and yellowish infarcts. There were large vegetations on the mitral valve, and many of the chordæ tendiniæ were destroyed by ulceration. *See Insp.* 1868, No. 50.

1509   Abscess of the Spleen.

An enlarged spleen laid open to shew an abscess-cavity occupying the lower three-quarters of the organ. The interior of the abscess is fibrous and trabeculated, and its wall is in great part formed by thickened splenic capsule. On the reverse of the specimen the upper extremity of the viscus is seen to be black, diffluent, and gangrenous.   *Presented by Dr. Bright.*

1510   Pyæmic Abscess of the Spleen.

A portion of a spleen shewing on its concave surface near its anterior margin a small cavity which in the recent state contained pus.

James J., æt. 37, was admitted under Mr. Cooper Forster, and died from pyæmia following amputation. At the autopsy abscesses were also found in the lungs and liver.

1511   Abscess of the Spleen opening into the Colon.

A spleen divided to shew at its upper part a portion of an abscess containing inspissated pus. A blue rod in-
. dicates a fistulous communication between the cavity and the adherent transverse colon.

Ann C., æt. 25, was admitted under Dr. Bright in 1825 for abdominal pain and frequent vomiting, having suffered 3 years previously from hæmatemesis. She had occasional rigors and her left hand became gangrenous. She died eleven days after admission, and at the autopsy the abscess was found to involve about half of the substance of the spleen and its contents had a "grumous chocolate appearance." There was a second abscess in the left ovary. *See Guy's Hosp. Reps.* 1838, p. 425.

1512   Perisplenic Abscess.

A spleen with a portion of the cardiac end of the stomach. The convex surface of the spleen is denuded of its capsule and covered by a layer of recent lymph, which extends over the adjacent wall of the stomach. In the recent state this roughened area formed part of the wall of an abscess-cavity, reaching to the left lobe of the liver.

Ann H., æt. 21, was admitted under Dr. Bright in 1829 with "frequent vomiting, under which she sank in about ten days, her legs having become œdematous." At the autopsy a small circular aperture, scaled with recent lymph, was found at the cardiac end of the stomach, and there were purulent thrombi in the iliac and femoral veins and in the inferior vena cava. *See Insp.* vol. 7, p. 121; and *Guy's Hosp. Reps.* 1838, p. 427.

1513   Gumma of the Spleen.

A portion of an enlarged spleen, the cut surface of which shews a mass of firm yellowish material extending from the capsule for a distance of two inches into the substance of the organ. Histologically the adventitious material has the structure of a gumma.

James S., æt. 41, was admitted under Mr. Cooper Forster with a strangulated hernia, for the relief of which herniotomy

was performed. He had suffered from syphilis, and was in a
"wretched cachectic condition." Two days after the operation
he died, and at the autopsy the liver was found to be cirrhotic;
the spleen weighed 44 ounces, and the testes were fibroid. *See
Insp.* 1861, No. 64.

1514   Gummata of the Spleen.

A portion of an enlarged spleen divided to shew deeply
embedded in its substance a hard yellow globular mass
about three-quarters of an inch in diameter. It is
separated by a fibrous investment from the surrounding
splenic tissue, and histologically has the structure of a
gumma. On the reverse of the specimen a similar
mass is seen situated close beneath the capsule of the
organ.

Francis W., æt. 38, was admitted under Mr. Bryant for
epileptiform fits associated with syphilitic necrosis of the skull.
He was trephined, and for a time his symptoms were relieved,
but subsequently he became hemiplegic, and died from pyæmia.
At the autopsy the liver and spleen were found to be adherent
to the diaphragm and to contain gummata. There was recent
pleurisy and peritonitis. *See Insp.* 1860, No. 105 ; and *Trans.
Path. Soc.* vol. 12, p. 11.

1515   Miliary Tuberculosis of the Spleen.

A portion of a spleen, the surface of which is granular
from the presence beneath the capsule of numerous
miliary tubercles. The section of the organ shews
similar nodules to be thickly scattered throughout its
substance. Histologically they are tubercles.

George B., æt. 6, was admitted under Dr. Rees with symptoms
of bronchitis, having 8 weeks previously suffered from measles.
He died two days after admission, and at the autopsy the bronchial
glands were found to be caseous and there were miliary tubercles
in the lungs, liver, and kidneys. *See Insp.* 1856, No. 132.

1516   Tubercle of the Spleen.

A small portion of a spleen, on the cut surface of which
are seen numerous firm yellow nodules varying from a
line to a quarter of an inch in diameter. A few similar

nodules project beneath the capsule and are covered by recent lymph. Histologically the nodules are seen to have the structure of tubercle.

From a negro who died of phthisis. Tubercles were also found in the liver. *See Old Museum Book*, No. 6.

Presented by Dr. Cholmeley.

1517 **Tuberculosis of the Spleen.**

A spleen partially injected and divided to shew its cut surface thickly beset with white nodules measuring about a line in diameter. Some of the larger nodules present a small central excavation, and histological examination shews them to have the structure of case-ating tubercle. *Presented by Sir Astley Cooper.*

1518 **Tuberculosis of the Spleen.**

A portion of an enlarged spleen, on the cut surface of which are seen numerous rounded and irregular-shaped areas of a yellowish colour and firm consistency. They occupy roughly one-half of the section, and similar nodules project beneath the capsule of the organ. Histologically they have the structure of caseous tubercle.

Albert G., æt. 6, was admitted under Dr. Washbourn for continued fever and enlargement of the spleen. He died nine weeks after the commencement of his illness, and at the autopsy a bronchial gland was found to be caseous, and there was tuberculous disease of the lungs, liver, intestines, and kidneys. The spleen weighed ten ounces and contained two infarcts. *See Insp.* 1893, No. 96.

1519 **Calcareous Nodules in the Spleen.**

Portions of a spleen in which are embedded two calcareous nodules, one of the size and shape of a mustard-seed, the other oval and measuring a quarter of an inch in its longest diameter.

1520   Spleen in Typhoid Fever.

A spleen uniformly enlarged and shewing fragments of
recent lymph upon its capsule.   Its weight is 7 ounces,
and it was of soft consistency.

> Henry H., æt. 13, was admitted under Mr. Howse with abdo-
> minal pain and distension, attributed to a blow received 3 weeks
> previously.   He was found to be suffering from typhoid fever,
> and 3 days after admission he died.   At the autopsy a perforation
> was discovered in the base of a typhoid ulcer ten inches above the
> ileo-cæcal valve.   There was acute suppurative peritonitis, and
> the peritoneal cavity contained 18 ounces of turbid fluid.   *See
> Insp.* 1803, No. 440.

1521   Leucocythæmic Spleen.

A spleen weighing 54 ounces, and measuring 11 inches
in length and 6 in breadth.   Its capsule is smooth and
free from adhesion.

> From a woman, æt. 40, who was admitted under Dr. Oldham
> for leucocythæmia.   Mr. Bryant removed the spleen, and 3 hours
> later the patient died.   At the autopsy more than a pint of
> blood was found in the peritoneal cavity in the region of the
> spleen.   The liver weighed 138 ounces.   *See Insp.* 1867, No. 283.

1522   Enlarged Spleen in Lymphadenoma.

The half of a uniformly enlarged spleen which measures
9 inches in length by 5½ inches in breadth, and in the
recent state weighed 46 ounces.   On the cut surface
are seen several pale infarcts, the largest extending
inwards 2 inches from the margin of the organ.   Elsewhere the section has a speckled appearance from the
presence of numerous rounded white nodules scattered
through the dark red spleen-pulp.   On the reverse of
the specimen the capsule shews slight thickening and
depressed white areas corresponding to the position of
the infarcts.   Histologically portions of the spleen are
seen to be more fibrous than usual and to contain an
excess of small round cells.

> Charles T., æt. 26, was admitted under Dr. Hale-White for
> anæmia, ascites, and œdema of the legs, having 8 months pre-

viously been treated in the hospital for Hodgkin's disease. On
admission a mass of enlarged glands was felt in the left iliac
fossa, and after six pints of ascitic fluid had been withdrawn
the spleen was found to be considerably enlarged. There was
irregular pyrexia, with diarrhœa and slight leucocytosis. The
patient died 4 days after admission, and at the autopsy the
lymphatic glands throughout the body were found to be enlarged,
and there was a slight degree of lardaceous change in the
kidneys, liver, and spleen. *See Insp.* 1891, No. 77; and *Trans.
Path. Soc.* vol. 42, p. 468.

1523   Spleen in Hodgkin's Disease.

A portion of a spleen with a mass of enlarged lymphatic
glands at its hilum. The surface of the organ is coarsely
nodular, and its section shews numerous white areas
which histologically are seen to consist of hyaline
fibrous material without any appearance of caseation.

Ellenborough K., æt. 10, was admitted under Dr. Bright in
1828 with enlargement of the spleen and cervical glands. Anasarca supervened, and the patient died about six weeks after
admission. At the autopsy the spleen was found to be "at least
four times its natural size," and the mediastinal and abdominal
lymphatic glands were large and exhibited "a firm cartilaginous
structure." *See Insp.* vol. 6, p. 156; and *Prep.* 1511 (12) [2nd
Edit.].

1524   Sarcoma of the Spleen.

A large spleen, part of the convex surface of which has
been removed to shew at its margin an encapsuled
mass of soft white growth measuring three inches in
diameter. Histologically it has the structure of a round-
celled sarcoma.

John F., æt. 30, was admitted under Mr. Key in 1821 for
paraplegia. At the autopsy growths were found in the spine,
pleura, and other parts of the body. *See Preps.* 396, 1359, and
[2nd Edit.] 1028, 1042, 1449, & 1544.

1525   Sarcoma of the Spleen.

A section through an enlarged spleen shewing .the
greater part of its substance to be replaced by rounded
masses of growth. On section some parts of the growth

are seen to be vascular, whilst others are of a whitish colour contrasting with the surrounding splenic tissue. On the reverse of the specimen several protuberant masses of growth are visible, one of which has perforated the capsule and was torn in removing the organ. Histologically the growth is a very vascular round-celled sarcoma.

Robert R., æt. 54, was admitted under Dr. Hale-White with a glandular swelling in the right axilla and enlargement of the liver, a tumour noticed for half a year having been removed from his forearm six months before his admission. A month later he died, and at the autopsy secondary deposits were found in the lungs, pericardium, peritoneum, and liver. *See Insp.* 1890, No. 329.

1526   **Melanotic Sarcoma of the Spleen.**

A spleen divided to shew embedded in its substance and projecting beneath its capsule a globular mass measuring two and a half inches in diameter. The cut surface shews the growth to be partially encapsuled and its substance to consist of a whitish material interspersed with nodules and irregularly-shaped masses of a dark reddish-brown colour. Histologically the growth is a very vascular melanotic sarcoma containing round and oval cells.

Anne O., æt. 40, was admitted under Dr. Moxon for ascites. She had previously suffered from pain in the back with frequent attacks of vomiting, and died one month after the onset of her illness. At the autopsy melanotic growth was found in the skin, pleura, peritoneum, jejunum, lungs, liver, and kidneys. *See Insp.* 1885, No. 99 ; and *Preps.* 933, and 1061.

1527   **Melanotic Sarcoma of the Spleen.**

A section of spleen, embedded in which are several black nodules of growth, the largest measuring $2\frac{1}{4}$ inches in diameter. Histologically the growth is a melanotic sarcoma containing large round pigmented cells.

From a young woman, who died shortly after amputation of the breast for malignant disease.

From the Fort Pitt Museum.

1528 Melanotic Sarcoma of the Spleen.

A spleen containing within its substance a large number of soft black nodules varying from a line to three-quarters of an inch in diameter. On the reverse of the specimen similar nodules are seen forming prominences beneath the capsule of the organ. Histologically the growth consists of round cells, most of which are pigmented.

John B., æt. 35, was admitted under Dr. Habershon for enlargement of the abdomen which had been noticed for six weeks. He died on the day of his admission, and at the autopsy melanotic growths were found in the stomach, liver, and many other parts of the body. *See Insp.* 1875, No. 230; and *Prep.* 717.

1529 Myeloid Sarcoma of the Spleen.

A portion of a spleen boldly projecting from the under surface of which is seen an oval mass of growth measuring two inches in its longest diameter. On section its margin is seen to be well defined, and its substance is markedly vascular. Histologically the growth is a myeloid sarcoma.

Susan G., æt. 45, was admitted under Dr. Pavy with rapidly growing lumps beneath the skin and bleeding from the gums. Her illness, which had lasted three weeks before admission, ended fatally a fortnight later with cerebral symptoms. At the autopsy growths similar to those in the spleen were found in the brain, lungs, and liver. *See Insp.* 1861, No. 62; and *Preps.* 333 & 1367.

1530 Carcinoma of the Spleen.

A small spleen containing a secondary deposit of growth of globular form and measuring two inches and a half in diameter. The mass is white in colour and friable,

its central portion having undergone softening and excavation.   Histologically it is a cylindrical-celled carcinoma.

Jane H., æt. 57, was admitted under Mr. Symonds for intestinal obstruction, due to malignant disease of the sigmoid flexure,
which was relieved by the removal of the growth and the esta-
blishment of an artificial anus.   An attempt was subsequently
made to close the opening, and a few days later the patient died
from suppurative peritonitis.   At the autopsy secondary deposits
were found in the liver and in the recto-vesical pouch of the peritoneal cavity.   *See Insp.* 1894, No. 343.

1531   Spleen invaded by Carcinoma.

A portion of a spleen shewing at its hilum a mass of
white growth invading the substance of the organ.
Histologically the growth has the structure of a spheroidal-celled carcinoma.

Ann B., æt. 35, was admitted under Dr. Addison in 1839
with an abdominal tumour apparently due to a malignant
growth in the omentum.   A month later she died, and at the
autopsy numerous malignant nodules and masses were found in
the peritoneum and liver.   The primary growth was thought
to be situated in the uterus.   *See Insp.* vol. 8, p. 59; and *Preps.*
2266 (48) & 2266 (54) [2nd Edit.].

1532   Cyst in the Spleen.

A portion of a spleen shewing at its upper end immediately beneath the capsule a thin-walled cyst about the
size of a walnut.   Its interior is trabeculated.   In the
recent state several cysts of the size of millet seeds were
noticed on other parts of the surface of the organ.

Jane D., æt. 54, was admitted under Dr. Habershon for
uræmic convulsions, and died four days later.   At the autopsy
the kidneys were found to be "extremely degenerated and covered
with cysts."   There was a small dermoid cyst in the left ovary.
See Insp. 1859, No. 65.

1533   Cyst in the Spleen.

A section of a spleen shewing upon its convex surface, just beneath the capsule, a cyst of about the size of a hazel-nut. The interior of the cavity is rugose on its splenic aspect, and its lining is formed of thin fibrous tissue.

Samuel B., æt. 31, was admitted under Dr. Wilks for phthisis, from which ten weeks later he died. At the autopsy the cyst in the spleen was found to contain a greenish, highly albuminous fluid. *See Insp.* 1883, No. 352.

1534   Cysts in the Spleen.

A spleen divided to shew at its upper end a congeries of thin smooth-walled cysts varying in size from an inch and a half to a quarter of an inch in diameter.

Charles B., æt. 45, was admitted under Dr. Back for ascites and phthisis, from which 18 days later he died. At the autopsy, in addition to tuberculous disease of the lungs, chronic peritonitis and perihepatitis were discovered, and the kidneys were granular. The cysts in the spleen contained fluid, but no hydatid membrane. *See Insp.* vol. 10, p. 11.

1535   Suppurating Hydatid of the Spleen.

An enlarged spleen from the inner surface of which projects an oval cyst about 5 inches in its longest diameter. An opening has been made in it to shew a thick fibrous wall and a number of adherent hydatid cysts similar to those seen at the bottom of the preparation-jar. In the recent state the sac contained purulent fluid and cheesy material.

Ada W., æt. 26, was admitted under Dr. Taylor with an abscess of the liver, which was drained through the pleura. At the autopsy the spleen was found to weigh 24 ounces, and the liver and lungs contained numerous small abscesses. *See Insp.* 1888, No. 381.

1536   **Suppurating Hydatid of the Spleen.**

A portion of an enlarged spleen shewing a rough lymph-covered surface which formed part of the wall of a suppurating hydatid cyst. The blue rod indicates a sinus traversing the spleen and opening externally in the left linea semilinearis about an inch below the margin of the ribs.

Henry S., æt. 39, was admitted under Dr. Habershon with a fluctuating swelling in the left hypochondriac region from which some months before hydatid fluid had been evacuated. Three months later 5 pints of turbid liquid containing hydatid cysts were removed through a large canula and permanent drainage established. The patient died about eight weeks after the operation, and at the autopsy the abscess in the spleen was found to communicate with the lesser sac of the omentum. *See Insp.* 1868, No. 99.

Section XXXI.—DISEASES OF THE SUPRARENAL BODIES.

1537 Accessory Suprarenal Body.

A fœtal suprarenal body with the kidney, shewing upon the former a nodule measuring about a line in diameter, and having the structure of the normal gland. It is attached to the kidney by a loose investment of areolar tissue. In describing this specimen Dr. Hodgkin observes : " It would seem that accessory bodies of this kind, though not invariably, are frequently present, and are liable to enlargement from disease."

1538 Accessory Suprarenal Bodies.

A kidney from which the capsule has been removed to expose on its anterior surface and at its upper end a thin plaque of a yellowish material which measures about $\frac{1}{20}$ of an inch in thickness and an inch and a half in its longest diameter. Histologically the plaque has the structure of the normal suprarenal body, being in some places continuous with the cortex of the kidney, and elsewhere separated from it by a thin layer of fibrous tissue.

Christopher K., æt. 22, was admitted under Dr. Hale White with lobar pneumonia and pericarditis, from which two days later he died. At the autopsy each kidney was found to have attached to it beneath the capsule a thin plate of suprarenal tissue in addition to a suprarenal body of normal shape, and in the normal position. *See Insp.* 1893, No. 29 ; and *Trans. Path. Soc.* vol. 45, p. 141.

1539   Hæmorrhage into the Suprarenal Body.

A left suprarenal body enlarged so as to be half the size of the kidney. The enlargement is produced by a considerable extravasation of blood into the medulla of the gland. In the recent state the cortical part of the capsule, though distended, appeared everywhere intact, and formed a complete investment to the coagulum within.

Edward H., æt. 41, was admitted under Dr. Gull in a moribund condition, suffering from pneumonia and hæmaturia, and died on the following day. At the autopsy the left lung was in a condition of grey hepatisation, and some of the smaller branches of the pulmonary artery were filled with thrombus. There was a renal calculus. *See Insp.* 1865, No. 66.

1540   Hæmorrhage into the Suprarenal Body.

A portion of a right suprarenal body which in the recent state was twice as large as the left and was infiltrated with blood. Histologically the vessels of the organ are seen to be filled with thrombus, and the glandular tissue is loaded with extravasated blood-corpuscles.

Maurice T., æt. 19, was admitted under Dr. Perry for pleurisy and pericarditis, having ten days previously suffered from inflammation of the ankles and knee-joints. Ten days later he died, and at the autopsy the heart weighed 21 ounces and its cavities were dilated. There were small soft vegetations upon the aortic and mitral valves. *See Insp.* 1894, No. 403.

1541   Lardaceous Suprarenal Bodies.

The suprarenal bodies, which are firmer than normal and have a translucent appearance, changes which on microscopical examination are seen to be due to a uniform infiltration of the organs with lardaceous material.

Emily R., æt. 28, was admitted under Dr. Rees for general anasarca and albuminuria, having previously suffered from syphilis. She died two months after admission, and at the autopsy the liver was found to be affected by perihepatitis. The liver, intestines, and kidneys were highly lardaceous. *See Insp.* 1869, No. 64; and *Prep.* 1270.

1542   Caseous Suprarenal Body.

A right suprarenal body with the kidney and adjacent lymphatic glands. The suprarenal body is greatly enlarged, and measures two inches in length and an inch and a quarter in breadth. On section it presents a uniform pale yellow colour, and histologically it is found to consist of caseous material with giant cells. The lymphatic glands are also caseous.

John S., æt. 42, was admitted under Dr. Barlow and died from carcinoma of the stomach. At the autopsy the apex of the right lung was indurated; the left suprarenal body was healthy. *See Insp.* 1860, No. 36.

1543   Caseous Suprarenal Bodies.

The suprarenal bodies, one of which is much smaller, and the other larger, than is normal. In both the glandular structure is replaced by fibrous and caseous material.

John B., æt. 36, was admitted under Dr. Rees in a condition of great prostration and almost pulseless. He had been getting weaker, and had suffered from occasional vomiting for about six months. After admission the vomiting became more frequent, and some signs of phthisis were detected. Addison's disease was suspected, but no discoloration of the skin was discovered. He died six days after admission, and at the autopsy the lungs shewed signs of old and recent phthisis, the other viscera being healthy. *See Insp.* 1866, No. 124.

1544   Suprarenal Body in Addison's Disease.

A right suprarenal body with the kidney. The former has been divided, and is seen to be converted into a mass of yellow material, which histologically is of a uniformly caseous character. In the recent state the suprarenal bodies were " at least four times their natural ₄thickness."

Anne R. was admitted under Mr. Key in 1829 for a tumour of the breast, and two days later was transferred to Dr. Cholmeley's care on account of severe constitutional symptoms. Her complexion was noticed to be very dark, and she was extremely feeble and emaciated. She frequently vomited, and becoming drowsy and delirious died on the day after her admission into the medical wards. At the autopsy the lungs presented old and recent tubercles, and an abscess in the left suprarenal body contained two drachms of pus. See Insp. vol. 14, p. 65; and Bright's Reports of Medical Cases, vol. 2, part 1, case 119, p. 247.

1545   Suprarenal Bodies in Addison's Disease.

The suprarenal bodies with the kidneys. The right suprarenal is much contracted, and is converted into dense fibrous tissue. The left, which in the recent state was as large as a hen's egg, presents a central abscess cavity with caseous contents and fibrous wall.

James W., æt. 32, was admitted under Dr. Golding Bird in 1850. For three years his skin had been becoming darker, and for one year he had suffered from extreme weakness with progressive anæmia. He died from pericarditis, and at the autopsy " no chronic disease was found except that of the suprarenal capsules." This is stated to be the first case in which any connection was thought to exist between the discoloration of the skin and the disease of the capsules. See Drawings 353 (5, 6, and 8); Wax Model; and Dr. Addison's Works, New Sydenham Soc. p. 217.

1546   Suprarenal Bodies in Addison's Disease.

The suprarenal bodies incised to shew their structure replaced by caseous and calcareous deposit.

Henry P., æt. 26, was admitted · under Dr. Addison for vertigo, head-ache, and occasional fainting fits, from which he had suffered for one month. For six months he had complained of pains in the back and legs, and for two or three months his face had been observed to be discoloured. On admission he was found to have an angular curvature of the spine, his complexion was of a dirty yellow tint, and the inside of his lips was stained nearly black. He died four weeks after admission, having suffered from constant hiccough during the last few days of his life. At the autopsy the 1st and 2nd lumbar vertebræ were found to be carious, and there was a large psoas abscess. The lungs contained tubercles. *See Insp.* 1854, No. 234; *Dr. Addison's Works, New Sydenham Soc.* p. 222; and *Drawings* 159 (67), § 353 (17).

1547 Suprarenal Body in Addison's Disease.

A suprarenal body which is atrophied, its proper structure being replaced by small caseous nodules embedded in a fibrous matrix.

Charles W., æt. 24, was admitted under Dr. Barlow in 1855 with pigmentation of the skin, more especially about the lower extremities, nausea, and asthœnia, these symptoms having been observed for five months.

1548 Suprarenal Bodies in Addison's Disease.

The suprarenal bodies with the kidneys and portions of the aorta and vena cava, mounted to shew the suprarenals converted into a mass of fibrocaseous material, in which under the microscope giant cells are seen.

Robert B., æt. 12, was admitted under Dr. Addison with discoloration of the skin, pain in the epigastrium, and frequent sickness. He died 8½ months after the onset of symptoms, and at the autopsy the whole of the body was of a brown hue except the palms of the hands and the soles of the feet. But for the suprarenal bodies the viscera were healthy. *See Insp.* 1859, No. 143; *Prep.* 1641 (10) [2nd Edition]; and *Wax Models* 358, 359.

1549 Suprarenal Bodies in Addison's Disease.

The suprarenal bodies with the kidneys and their vessels mounted to shew the suprarenals somewhat enlarged and converted into a yellow fibrocaseous material embedded in which are white cretaceous deposits.

George A., æt. 23, was admitted under Dr. Habershon in a prostrate and depressed condition with excessive pigmentation of the skin. The face was of an olive-brown colour and the genital organs very dark, the discoloration elsewhere being but slight. He sank from exhaustion ten days after admission, and at the autopsy the apex of the right lung presented a fibrous nodule with a cretaceous centre. *See Insp.* 1861, No. 75.

1550   Suprarenal Body in Addison's Disease.

A left suprarenal body with the kidney. The former has been incised, and its cut surface exhibits a dense grey substance, embedded in which are irregular masses of yellow cheesy material.

From Edward P., æt. 32, who four years before his death was seized with vomiting and purging which continued for some time, being followed by a constant sinking feeling in the epigastrium. His skin from the beginning of his illness had a yellowish tint, which increased in intensity until it became of the colour of mahogany. His death was preceded for 48 hours by constant sickness with extreme prostration. At the autopsy no trace of the right suprarenal capsule was found. The other organs were healthy, and there was abundant subcutaneous fat. *See Guy's Hosp. Reps.* 1862, p. 59.

Presented by Dr. Stedman.

1551   Suprarenal Bodies in Addison's Disease.

A dissection displaying the suprarenal bodies and their connections with the semilunar ganglia. The glands are of unequal size, one being larger and the other smaller than normal. Both are disorganized by disease, being converted into a yellow material which under the microscope is found to consist of a caseous deposit containing giant cells.

Edward G., æt. 18, was admitted under Dr. Habershon in 1863 for Addison's disease, from which he died about two years after the onset of symptoms. At the autopsy a sinus was found in the neck leading to scrofulous lymphatic glands, and there was a tuberculous nodule in one of the lungs. *See Guy's Hosp. Reps.* 1864, p. 82.

1552   Suprarenal Body in Addison's Disease.

A suprarenal body with the kidney. The suprarenal

body is disorganised, and on section is seen to consist of a tough fibrous investment surrounding a caseous mass, part of which has undergone softening.

From Harriet R., æt. 19, who was suddenly seized with pain in the head, became comatose, and died two and a half days after the onset of acute symptoms. Two years and seven months before her fatal illness she had been a patient under Dr. Habershon in Guy's Hospital, having then for about two years suffered from occasional vomiting and gradually increasing weakness. While in the hospital she had the appearance of a mulatto. At the autopsy the skin was almost as dark as that of a negress, and the subcutaneous fat was abundant.

Presented by Dr. Charles Webb, 1864.

1553   Suprarenal Body in Addison's Disease.

A suprarenal body in the centre of which is an irregular excavation which in the recent state contained a table-spoonful of pus. The walls of the cavity are ragged, and are composed of caseous material with an outer investment of tough fibrous tissue.

From a man, æt. 26, who, while under treatment for some febrile attack eighteen months before his death, was noticed to have patches of slight bronzing upon his skin. The bronzing gradually increased, and the only other symptoms from which he suffered were loss of appetite and muscular debility. He died from exhaustion with delirium. At the autopsy there was half an inch of fat upon the abdomen. " There was not a tubercle or scrofulous deposit to be seen, the body was as healthy as that of any young person killed by accident." *See Guy's Hosp. Reps.* 1865, p. 35.

1554   Suprarenal Bodies in Addison's Disease.

The suprarenal bodies somewhat contracted and exhibiting on their cut surfaces areas of caseous and cretaceous material embedded in the glandular tissue of the organs.

From a woman, æt. 50, who died in 1865 after 5 days incessant vomiting with great prostration, having suffered during the preceding six weeks from two similar attacks, each of three days' duration. A dingy-brown discoloration of the skin had been noticed for two or three months before her death. *See Guy's Hosp. Reps.* 1865, p. 36.

Presented by Dr. Gilbertson.

1555   Addison's Disease without Melasma.

The suprarenal bodies and kidneys, the former greatly enlarged and being about two thirds the size of the kidneys. Both pairs of organs have been divided, and the cut surface of the suprarenal bodies presents a uniform yellow appearance. Histologically they are seen to be converted into an amorphous caseous material. The neighbouring lymphatic glands and other structures are firmly united to the caseous masses, and the vena cava is seen to be invaded and perforated.

William T., æt. 31, was admitted under Dr. Gull for extreme weakness. He had been ailing for about four months, his chief symptoms being pain in the back and occasional vomiting. He died two days after admission. At the autopsy the complexion was noticed to be of a sallow tint, " such as is seen in cachetic persons." No increased pigmentation could be detected, nor had any change in the colour of his skin attracted attention during life. The viscera, with the exception of the suprarenal bodies, were normal. *See Insp.* 1862, No. 258.

1556   Addison's Disease with general Tuberculosis.

The parts concerned in Addison's disease dissected to shew the relation of the suprarenal bodies to the surrounding nerves and ganglia. The capsules are of unequal size, the right being about twice as large as normal and the left somewhat shrunken. The cut surface shews a caseous material with partial calcareous degeneration.

William H., æt. 17, was admitted under Dr. Fagge for debility and emaciation, having been in ill-health for one year. For three days before he came into the hospital he suffered from vomiting and diarrhœa. Subsequently he became delirious, and died comatose six days after admission. At the autopsy the neck, face, and right nipple presented a brown discoloration and the lungs liver, spleen, and kidneys contained numerous tubercles. *See Insp.* 1832, No. 166.

1557 Addison's Disease with old Spinal Caries.

The suprarenal bodies atrophied and converted into small masses of caseous material. Above is mounted a portion of the vertebral column shewing ankylosis of two adjacent vertebræ with destruction of the intervertebral disc, the results of spinal caries.

Margaret V., æt. 45, was admitted under Dr. Goodhart for vomiting and weakness. Some years previously she had suffered from a spinal abscess. She was noticed to have some patches of pigmentation about the cheeks and neck. Six months after admission she died, and at the autopsy the apices of the lungs presented evidences of healed phthisis. There was a calcareous mesenteric gland. *See Insp.* 1886, No. 407.

1558 Adenoma of the Suprarenal Body.

A portion of a right suprarenal body on the cut surface of which is seen, contained within the fibrous investment of the organ, a well-defined yellow nodule, which histologically is found to have the normal structure of the gland infiltrated with fat.

Thomas M., æt. 84, was admitted under Mr. Cock for gangrene of the leg and atheromatous arteries, aud died from pneumonia.

1559 Adenoma of the Suprarenal Body.

A suprarenal body containing a soft oval mass measuring an inch and a quarter in its longest diameter. In the recent state it was of a pink colour and appeared to be fatty. Histologically it consists of normal suprarenal tissue.

Hannah L., æt. 64, was admitted under Dr. Gull for heart-disease, from which she died. *See Insp.* 1864, No. 30.

ı 2

1560   Adenoma of the Suprarenal Body.

A transverse section through a suprarenal body, in the substance of which is a partially encapsuled ovoid tumour measuring half an inch. in length and a third of an inch in thickness. Histologically it has the normal structure of the gland.

Martha N., æt. 49, was admitted under Dr. Pavy for obesity and dropsy, and died from fatty degeneration of the heart. At the autopsy a similar adenoma was found in the other suprarenal body. *See Insp.* 1869, No. 256.

1561   Adenoma of the Suprarenal Body.

A right suprarenal body, from the upper surface of which projects a smooth ovoid tumour measuring two and a quarter inches in its longest diameter. The tumour is covered by the delicate fibrous investment of the suprarenal body, and its cut surface presents the appearance of the cortical substance of the organ. Histologically the growth has the structure of the suprarenal capsule infiltrated with oil globules.

Mary D., æt. 57, was admitted under Mr. Davies-Colley for carcinoma of the uterus. At the autopsy there was found to be a hydronephrosis on the left side. The left suprarenal body was normal. *See Insp.* 1889, No. 141.

1562   Carcinoma of the Suprarenal Body.

A half of a suprarenal body, the central part of which is occupied by an ovoid growth measuring two inches in its longest diameter. The peripheral portion of the tumour consists of a soft white material, while its centre is composed almost entirely of extravasated blood. Histologically the growth has the structure of a cylindrical-celled carcinoma.

John D., æt. 72, was admitted under Dr. Bright in 1826 with symptoms of phthisis and dyspepsia. At the autopsy the lungs were found to be tuberculous, and there was a large carcinomatous ulcer in the stomach. There were secondary deposits of growth in the pleura and peritoneum. *See Red Insp. Book*, p. 166; and *Prep.* 680.

1563   Carcinoma of the Suprarenal Bodies.

The halves of two suprarenal bodies greatly enlarged and almost entirely destroyed by a soft white growth, which histologically has the structure of a cylindrical-celled carcinoma. In some parts a thin shell of suprarenal tissue is discernible at the periphery of the tumour.

John R., æt. 54, was admitted under Mr. Cock with partial paraplegia, at first attributed to injury. Twenty-one days later he died, and at the autopsy a large carcinomatous tumour was found in one of the kidneys, and there were secondary deposits in the dura mater, liver, and in the gastric and lumbar glands. *See Insp.* 1863, No. 271.

1564   Carcinoma of the Suprarenal Body.

A left suprarenal body with the kidney mounted to shew the former, which is somewhat enlarged, infiltrated and destroyed by a firm brownish growth having the histological characters of a scirrhous spheroidal-celled carcinoma.

Elizabeth L., æt. 53, was admitted under Dr. Babington in 1853 for carcinomatous disease of the stomach, from which, two days later, she died. At the autopsy the right suprarenal body was normal, and there were secondary deposits in the mediastinal and abdominal lymphatic glands. *See Insp.* vol. 37, p. 104; *Drawings*, 150 (68 & 69) & 353 (16); and *Dr. Addison's Works, New Sydenham Soc.* p. 234.

1565   Carcinoma of the Suprarenal Body.

A right suprarenal body, enlarged so as to be two thirds of the size of the kidney to which it is attached. It has been divided, and is seen to be replaced by a whitish almost homogeneous material, which histologically has the structure of a spheroidal-celled carcinoma with scanty stroma.

William G., æt. 27, was admitted under Dr. Barlow with the symptoms of mediastinal tumour, from which ten days later he died. At the autopsy a large cancerous mass occupied the left side of the thorax, invading the ribs. The left suprarenal body was normal. *See Insp.* 1855, No. 14; *Prep.* 1050 (40) [2nd Edit.], *Drawing,* 352 (12); and *Dr. Addison's Works, New Sydenham Soc.* p. 238.

1566   Carcinoma of the Suprarenal Body.

A left suprarenal body laid open to shew one half of it enlarged and occupied by a white mass of adventitious deposit, clearly defined from the darker glandular tissue. Histologically the growth has the structure of a spheroidal-celled carcinoma with scanty stroma.

George R., æt. 59, was admitted under Dr. Pye-Smith, and died, comatose, six days later. At the autopsy the liver was found to be cirrhotic, and to contain carcinomatous deposit. Similar deposits were found in the kidney, pancreas, thyroid gland, stomach, and pericardium, and also in the glands of the mesentery and lesser omentum. *See Insp.* 1892, No. 101 ; and *Preps.* 109 & 689.

1567   Carcinoma of the Suprarenal Body.

The half of a left suprarenal body expanded to enclose a mass of growth measuring three inches in its longest diameter, and having the histological characters of a carcinoma.

Catherine F., æt. 42, had been an out-patient for some time under Mr. Callaway's care, suffering from an enlarged thyroid gland. She was subsequently admitted into the hospital and died. At the autopsy the thyroid gland was found to be affected by a hard carcinomatous growth, and there were secondary deposits in the lung. *See Insp.* vol. 32, p. 84 ; and *Prep.* 108.

1568   Carcinoma of the Suprarenal Body.

The half of a tumour occupying the position of the left suprarenal body and loosely attached to the upper end of the kidney. It has an irregular ovoid form, and measures 5 inches in its longest diameter. The cut surface shews it to consist of a soft dark-coloured deposit traversed by numerous fibrous trabeculæ. Its outer surface is smooth and lobulated, and histologically its structure is that of a carcinoma.

Stephen B., æt. 29, was admitted under Dr. Addison in 1842 with pain in the left lumbar region of two months' duration. A tumour was felt below the margin of the liver, and was found to be solid after exploratory puncture. The patient died 7 weeks after admission, and at the autopsy a large soft growth occupied nearly the whole of the right lobe of the liver, and there were secondary deposits in the lungs. *See Insp.* vol. 32, p. 83.

1569   Sarcoma of the Suprarenal Bodies.

The suprarenal bodies attached to the kidneys, which they greatly exceed in size. The enlargement is seen to be due to a malignant deposit, consisting of nodular growths varying in size from a pea to a tangerine orange. These nodules are united to each other by fibrous tissue, and histologically have the structure of a very vascular small round-celled sarcoma.

Frederick B., æt. 5 months, was admitted under Dr. Pye-Smith with swellings on the head, first noticed at the age of 2 months. He died a month after admission, and at the autopsy sarcomatous growth was found in the cranium, mediastinum, liver, ribs, and in the subcutaneous tissue of the arm and leg. *See Insp.* 1881, No. 289.

1570   Sarcoma of the Suprarenal Body.

A right suprarenal body divided to shew enclosed within its cortex a nodule of soft material about as large as a hazel nut. In the recent state the growth was in parts white and in parts red from the presence of blood. Histologically it has the structure of a small round-celled sarcoma.

Agnes M., æt. 8 months, was admitted under Dr. Perry with marasmus, and ten days later died from bronchopneumonia. *See Insp.* 1894, No. 334.

1571   Sarcoma of the Suprarenal Body.

A right kidney, the suprarenal body of which is replaced by a large ovoid tumour measuring two inches and a quarter in its longest diameter. The growth has been incised, and is seen to be composed of a white material infiltrated in most parts with blood. Histologically the growth is a highly vascular small round-celled sarcoma.

Henry C., æt. 10 months, was admitted under Dr. Pitt with a tumour in the right hypochondrium which had been noticed for 9 days. The tumour increased in size, jaundice supervened, and the patient died one month after admission. At the autopsy the liver was found to be occupied by a large growth, and there were secondary deposits in the glands at the head of the pancreas. *See Insp.* 1895, No. 388.

1572   Suprarenal Body in Mycosis Fungoides.

A right suprarenal body, uniformly enlarged and weighing 4½ ounces. The gland is infiltrated and replaced by a soft deposit, which in the recent state was vascular and in parts yellow and caseous. Histologically the growth consists of an aggregation of small round cells with numerous necrotic areas.

John B., æt. 66, was admitted under Dr. Pye-Smith for general dermatitis associated with multiple tumours of the skin, some of which sloughed. He died from gangrenous broncho-pneumonia. *See Insp.* 1891, No. 107; *Prep.* 1657 (11) [2nd Edit.]; and *Drawings.*

Section XXXII.—INJURIES AND DISEASES OF THE KIDNEYS AND URETERS.

Hydatid of the Kidney: 1679-1683.
Hydronephrosis: 1684-1703.
Dilated Ureters: 1704-1709.
Stenosis of Ureter: 1710-1712.
Pyonephrosis: 1713-1718.
Calculi in the Kidney: 1719-1727.
Calculi from the Kidney: 1728-1737.
Thrombosis of Renal Vessels: 1738-1740.
Atheroma of Renal Arteries: 1741.

1573   Persistent Lobulation of the Kidney.

The halves of two kidneys of adult size, mounted to
shew the persistence of the lobulation which normally
exists in the fœtal organ.

1574   Single Kidney.

A solitary kidney, weighing about six ounces, which
occupied the usual situation on the right side. It has
been laid open and has a normal structure. Below is
mounted a portion of the bladder in which is seen the
opening of the right ureter, the left being absent.

> James R., æt. 49, was admitted under Dr. Pavy for hæmo-
> ptysis, and died from hæmorrhage into the pleural cavity. At the
> autopsy the blood was found to have proceeded from the ruptured
> sac of an aortic aneurism. No trace of the left kidney or supra-
> renal body was discovered. *See Insp.* 1874, No. 384.

1575   Single Kidney.   Hydronephrosis.

A single kidney from the right side, considerably
enlarged, and laid open to shew its pelvis to be somewhat dilated. In the recent state the pelvis formed a
tense sac, from which the urine could not be forced
into the ureter until the tension of the sac had been
diminished by partial evacuation of its contents.

> William K., æt. 28, was admitted under Dr. Goodhart for
> persistent vomiting and hiccough of two weeks' duration, associated
> with diarrhœa and scanty micturition. A tumour was felt on the
> right side of the abdomen, and on the day following admission, no

urine having been passed, four ounces were withdrawn by catheter.
The patient became comatose and died the same evening. At the
autopsy the retro-peritoneal tissue on the right side was infiltrated
with fluid. No trace of the left kidney or ureter could be found.
The bladder was empty. *See Insp.* 1894, No. 107.

1576   Solitary Kidney.   Nephrotomy.

The half of a solitary kidney. The organ, which was
situated on the left side, is enlarged, and in the recent
state weighed fourteen and a half ounces. The
structure appears to be normal, and on the reverse of
the specimen is seen a ligature, around which the
capsule is thickened and partially coated with lymph.

Ellen L., æt. 36, was admitted under Dr. Taylor for pain and
swelling in the left lumbar region. The kidney was explored and
no calculus detected. A fortnight later the patient died from
pyæmia, and at the autopsy a perinephric abscess was found at the
seat of operation. No trace of the right kidney could be discovered. *See Insp.* 1890, No. 82.

1577   Horse-shoe Kidney.

Two kidneys united at their lower extremities by a
broad thin band, consisting chiefly of fibrous tissue.
The ureters pass downwards in front of the kidneys, and
the left is surrounded and constricted by a sheath of
firm white material, having the histological characters
of scirrhous spheroidal carcinoma.

From a sailor, æt. 66, who died from infiltrating carcinoma of
the stomach.

Presented by Mr. Hardy, Junr.

1578   Horse-shoe Kidney.   Hydronephrosis.

A horse-shoe kidney, the posterior half of which has
been removed to shew the organ in a condition of
hydronephrosis. Numerous black calculi of oxalate of
lime occupy the dilated calices. On the reverse of the
specimen is seen a mass of fibro-fatty material closely
adherent to the capsule of the kidney. The two halves

of the organ are united at the lower end by a considerable bridge of renal tissue.

James R., æt. 45, was admitted under Dr. Goodhart with hemiplegia affecting the left side. He became delirious, and died six weeks after admission. At the autopsy the heart was found to weigh eighteen and a half ounces, and there was recent pericarditis. Only one ureter was discovered, which arose from the left half of the kidney. *See Insp.* 1893, No. 214.

1579   Horse-shoe Kidney.   Nephrectomy.

A dissection of the kidneys and the adjacent parts from a case in which the left kidney (now replaced) had been removed by operation. At the lower end of the right kidney is seen a stump of renal tissue, by which that organ was connected with its fellow. The right kidney is enlarged, and its hilum is situated in front, its anterior surface presenting a broad groove running downwards and outwards, occupied by the vessels and ureter. The left kidney is small, and its anterior half has been removed to shew a condition of extreme hydronephrosis. The aortic glands are swollen from inflammatory deposit.

Francis F., æt. 28, was admitted under Mr. Lucas for pain in the left lumbar region and slight swelling on the corresponding side of the abdomen. There was pus in the urine. Nephrectomy was performed, and the patient died on the following day. At the autopsy the lungs were found to be affected by phthisis. *See Insp.* 1895, No. 450.

1580   Malposition of the Kidney.

A portion of the trunk of a fœtus dissected to shew the right kidney lying upon the vertebral column at the brim of the pelvis. The ureter is seen to proceed from the anterior aspect of the organ. The left kidney and suprarenal body occupy the normal situation.

1581   Misplaced Kidney.

A dissection of the kidneys with the great vessels and the left suprarenal body. The left kidney, which is of

ovoid shape, is seen to be displaced downwards, and in
the recent state lay in front of and below the promon-
tory of the sacrum. Its artery arises from the aorta at
its bifurcation, and its vein runs into the right internal
iliac vein. The ureter was short, and proceeded directly
from the hilum of the kidney to the bladder. The
right kidney is in the usual situation, and is provided
with two arteries. The left suprarenal body is not
displaced, and has a normal vascular supply.

Clara D., æt. 39, was admitted under Dr. Pavy for pernicious
anæmia, from which she died. *See Insp.* 1884, No. 35.

1582   **Malformation of the Kidney.**

A left kidney, which in the recent state lay upon the
vertebræ in a position lower than usual. There are
several grooves upon the anterior surface of the organ
where it is crossed by the vessels and ureter, converging
to the hilum, which is situated in front.

Edward R., æt. 50, was admitted under Dr. Bright in 1830,
and died from heart-disease and infarction of the lungs. *See
Insp.* vol. 9, p. 72.

1583   **Malformation of the Kidneys.**

A dissection of the kidneys with their vessels, the
aorta, and the inferior vena cava. In each kidney the
hilum is in front. The right kidney is displaced
downwards, and in the recent state lay upon the psoas
muscle at the brim of the pelvis. One artery arises
from the front of the aorta a little above its bifurcation,
and another from the common iliac at its termination.
The left kidney occupied its normal position, and is
united to its fellow by a thin fibrous band.

Emily T., æt. 35, was admitted under Dr. Hicks for pelvic
cellulitis following parturition. Ten days after admission she
died, and at the autopsy a mass of soft vegetations was found upon
one of the aortic valves. *See Insp.* 1880, No. 402.

1584   Mis-shapen Kidney.

The half of a healthy kidney, the transverse diameter of which is increased so as to give the organ a semilunar shape.

Richard D., æt. 40, was admitted under Dr. Addison for phthisis, from which he died. *See Insp.* 1859, No. 27.

1585   Deformed Kidney.

A kidney, the upper third of which is separated from the rest of the organ by a deep depression. A section has been made, and shews that at this situation the renal parenchyma has disappeared, and is replaced by a connecting band of fibrous tissue. On the surface of the lower portion of the kidney is seen a shallow depression with a puckered margin.

George S., æt. 58, was admitted under Dr. Barlow with dropsy and heart-disease, from which three weeks later he died. At the autopsy the mitral valve was found to be thickened, and there were vegetations upon its edges and upon the contiguous surface of the auricle. *See Insp.* 1860, No. 168.

1586   Kidney Deformed by Compression.

A left kidney considerably altered in shape as the result of pressure. Between the hilum and the convex border it is flattened out so that its transverse diameter is no greater than the usual thickness of the organ.

Abraham H., æt. 35, was admitted under Dr. Bright in 1828 with extreme deformity of the spine, and died from general tuberculosis. At the autopsy an abscess was found just below the diaphragm associated with caries of the subjacent vertebræ. *See Insp.* vol. 6, p. 49; and *Prep.* 1026 (50) [2nd Edit.].

1587   Kidney with two Pelves.

A child's kidney with two pelves, each provided with a ureter. The two ureters opened close together by separate orifices into the bladder.

1588   Malposition of the Pelvis of the Kidney.

A left kidney, the structures in the hilum of which have been partially dissected to shew the pelvis and ureter lying in front of the renal vessels.

Presented by Dr. Fagge, 1865.

1589   Kidney with two Ureters.

A left kidney, from which proceed two ureters, one of normal calibre which arises from the pelvis and enters the bladder in the usual situation. The other is tortuous and dilated so as to freely admit the middle finger. The dilated ureter proceeds from the upper end of the kidney and opens by an orifice about an eighth of an inch in diameter at the neck of the bladder in the middle line.

Louisa L., æt. 32, was admitted under Dr. Moxon for Bright's disease, and died six days later from uræmia. At the autopsy the kidneys were found to be in a condition of chronic parenchymatous nephritis. The right pelvis and ureter were normal. *See Insp.* 1883, No. 82.

1590   Kidney with two Ureters.

A left kidney, from the pelvis of which arise two ureters each appearing to be about as large as a normal ureter. They are united at their lower end about an inch above the point of entrance into the bladder.

1591   Kidney with two Ureters.

A right kidney with two ureters of equal size which open into the bladder by separate orifices divided by a narrow fold of mucous membrane.

1592   Kidney stained by Nitrate of Silver.

A small portion of kidney presenting a black discoloration of its pyramids and of its glomeruli from the internal administration of nitrate of silver. Histological examination shews that the metal is deposited in the form of minute granules in the malpighian bodies and in and between the tubules of the pyramids.   Below is mounted a portion of liver similarly affected.

From a man who died of epilepsy, for the relief of which he had for some time taken nitrate of silver.

Presented by Mr. E. Pye-Smith, 1859.

1593   Urates in the Kidney.

The half of a kidney injected and shewing upon its cut surface a yellowish-white deposit of amorphous urates arranged in streaks corresponding with the tubes of the pyramids.

Sarah S., æt. 40, was admitted under Dr. Wilks for œdema of the legs and dyspnœa, and died the following day.   At the autopsy the heart was found to be greatly hypertrophied, and the kidneys, which were granular on the surface, contained a few small cysts. *See Insp.* 1873, No. 237.

1594   Lacerated Kidney.

A left kidney from which the capsule has been stripped to shew upon the anterior surface several transverse lacerations passing across its upper end.   On the reverse of the specimen some of these lacerations are seen to be prolonged obliquely downwards towards the hilum of the organ.

From an adult male patient who was killed by a waggon-load of earth passing across his body.   At the autopsy all the ribs on the left side were broken, and a considerable quantity of blood was found in the pleural and peritoneal cavities.   The spleen was severely lacerated, one of its fragments being completely separated. *See Insp.* 1860, No. 158.

1595   Lacerated Kidney.   Perinephric Abscess.

A left kidney seen from behind lying exposed in the sac of a large abscess, the walls of which are formed by the perinephric tissue.   The organ is divided into equal

halves by a complete transverse laceration which involves the pelvis of the kidney. A red rod has been
passed into the renal artery, on one of the main branches
of which is seen an aneurysmal sac about a quarter of
an inch in diameter. A blue rod indicates the renal
vein which is ruptured. The communication between
the abscess cavity and the ureter is marked by a yellow
rod.

Edwin D., an adult, was admitted under Mr. Hilton having
been struck by the buffer of a railway engine. The next day he
passed urine which contained blood and casts of the ureters.
Hæmaturia persisted, and five weeks after the injury the patient
died with symptoms of peritonitis. At the autopsy it was found
that the peritoneum over the perinephric abscess had sloughed.
See Insp. 1866, No. 205.

1596   Lacerated Ureter.

A right kidney with the upper part of its ureter and a
portion of the inferior vena cava. There is a complete
transverse laceration of the ureter at its junction with
the pelvis of the kidney.

Emily N., æt. 33, was admitted under Mr. Poland for injuries
sustained by being crushed between a railway carriage and the
platform. Six days later she died, and at the autopsy the processes of the lumbar vertebræ were found to be fractured and the
inferior vena cava and the renal vessels contained thrombus. *See
Insp.* 1868, No. 25 A; and *Prep.* 1740.

1597   Lacerated Ureter.   Urinary Cyst.

A right kidney seen from behind, together with the
ureter and the retroperitoneal tissue, forming part of
the wall of a sac which in the recent state contained
three pints of urine. There is a solution of continuity
between the upper and lower parts of the ureter, into
the latter of which a yellow rod has been passed.

John P., æt. 6, was run over by a light cart, the wheel of
which crossed his abdomen. The next day he passed blood in his

urine. Three weeks later he died in convulsions, a large cyst having been felt in the right hypochondriac region a few days before his death.

Presented by Mr. Samuel Brown, 1865.

1598   Atrophied Kidney.

A left kidney, the capsule of which is thickened and firmly embedded in the perinephric fat, similar fatty material being present in excessive amount in the pelvis of the organ. The kidney measures only two inches in length, and portions of it examined under the microscope consist of fibro-fatty tissue enclosing relics of renal parenchyma.

Robert W., æt. 40, was admitted under Dr. Goodhart for pyrexia of some weeks' duration, and died from some obscure septic condition probably dependent upon tertiary syphilis. At the autopsy the right kidney showed compensatory hypertrophy. *See Insp.* 1895, No. 65.

1599   Atrophied Kidney.

An aorta and vena cava with the suprarenal and renal vessels of the right side. The adrenal body appears normal, and below are seen two small sacs measuring half an inch and an inch in diameter representing the remains of the right kidney.

Mary T., æt. 22, was admitted under Mr. Birkett for disease of the left knee-joint, and died after amputation through the thigh. At the autopsy a wasted ureter was found passing to the right kidney. *See Insp.* 1863, No. 9.

1600   Atrophied Kidney.

A left kidney greatly diminished in size, measuring only an inch in length and uniformly atrophied. The corresponding suprarenal capsule is of the ordinary size. Histologically the renal structure is almost entirely replaced by fibrous tissue.

Mary M., æt. 50, was admitted under Dr. Cholmeley in 1829 for intestinal obstruction. A year previously her uterus had been removed by Dr. Blundell for carcinoma. She died five days after admission, and at the autopsy deposits of secondary growth were found in the peritoneum, and the pelvic viscera were matted together. The right kidney was of the "usual size and healthy." *See Insp.* vol. 7, p. 137.

1601   **Localized Atrophy of the Kidney.**

A right kidney, weighing one ounce and a quarter, divided to shew its upper two thirds shrunken and fibrous. Over this part the capsule is thickened and firmly adherent. Where the capsule has been detached it shews beneath it an irregular nodulated surface. In the recent state several small calculi were lodged in the calices of this portion of the kidney. Histologically the tubular structure is replaced by fibrous tissue, and the glomeruli, which are mostly hyaline, occupy about a half of the cortical area.

James V., æt. 47, was admitted under Dr. Washbourn for phthisis, from which, four and a half months later, he died. At the autopsy tuberculous ulceration was found in the larynx and intestines. The left kidney weighed 8 ounces, and shewed compensatory hypertrophy. *See Insp.* 1892, No. 102.

1602   **Atrophy of the Kidney from Compression.**

A right kidney seen from behind, the upper part of its ureter being dilated, whilst the organ itself is considerably atrophied and hydronephrotic. On the reverse of the specimen is seen a large hydatid sac about six inches in diameter closely united to the remains of the kidney, which histologically exhibits an advanced condition of sclerosis.

Thomas L., æt. 22, was admitted under Dr. Bright in 1836 for a swelling in the abdomen. Eleven days later he died, and at the autopsy the peritoneal cavity was found to be almost completely occupied by hydatid cysts. Similar cysts were found in the liver and spleen. *See Insp.* vol. 23, p. 8.

1603   **Infarct of the Kidney.**

A portion of a kidney mounted to shew upon its sur-

K 2

face a well-defined depressed area, which on section is
seen to form the base of a wedge-shaped infarct. Histological examination shews that the details of the renal
structure in the infarcted area are ill-defined and stain
less readily than the surrounding tissue.

Alfred C., æt. 15, was admitted under Dr. Habershon for
chronic bronchitis and emphysema. Fourteen days later he
died, and at the autopsy the bronchial tubes were found to be
greatly enlarged, and the right side of the heart was hypertrophied
and dilated. The pulmonary artery was as thick as the aorta.
See Insp. 1874, No. 127.

1604   Acute Nephritis.

The halves of two kidneys greatly enlarged and marked
by persistent fœtal lobulation. The surface is smooth,
and in the recent state the cortex was pale and the
pyramids were congested.

Walter B., æt. 3, was admitted under Dr. Goodhart with
general anasarca of two months' duration. Whilst in the hospital
the cardiac impulse was noticed to move outwards three quarters
of an inch beyond the nipple line. Albuminuric retinitis was
detected four days before his death, which occurred eight weeks
after admission. The scrotum became inflamed and gangrenous.
At the autopsy the left ventricle was found to be hypertrophied,
and the kidneys weighed thirteen ounces, being four times the
normal weight at this age. *See Insp.* 1894, No. 30.

1605   Acute Nephritis.

The half of a kidney partially injected and somewhat
enlarged. Its surface is smooth, and in the recent state
the organ was " of a very pale or nearly white colour."
Histologically the epithelium of the tubes is swollen
and granular, and there is an excess of small round
cells in the connective tissue. This is one of Dr.
Bright's original preparations.

Edward M., æt. 25, was admitted under Dr. Cholmeley in
1827 with general anasarca attributed to exposure at sea three or
four months previously. He died ten days after admission, and
at the autopsy the pleural and peritoneal cavities were found to
contain large quantities of serous effusion. *See Insp.* vol. 4, p. 114.

1606 Tubal Nephritis.

The half of a kidney which is injected, and in the recent state presented a speckled appearance resulting from fatty degeneration of the tubal epithelium.

Mary S., æt. 25, was admitted under Dr. Bright in 1825 for œdema of the legs of two months' duration. The urine was found to be albuminous and to have a specific gravity of 1014. While in the hospital she suffered from urgent dyspnœa and frequent diarrhœa. She died two months after admission, and at the autopsy tuberculous excavation was found in the apex of the right lung, and there was considerable serous effusion into the pleural and peritoneal cavities. In describing the kidney Dr. Bright says :—"Internally the whole cortical structure was of a pretty uniform yellowish colour, with many small opaque and indistinct yellow spots." *See Insp.* vol. 2, p. 28 ; and *Bright's Medical Reports*, Part I. p. 12, pl. 2.

1607 Chronic Tubal Nephritis.

A kidney which is enlarged and has been laid open to shew the pyramids to be congested. The surface of the organ is smooth. Histologically the tubules are seen to contain numerous hyaline casts, the epithelial cells being granular, ill-formed, and many of them without visible nuclei.

Henry J., æt. 22, was admitted under Dr. Gull for general anasarca, for the relief of which his legs were punctured. Cellulitis supervened, and he died about 3 weeks after admission. The patient is stated to have suffered from dropsy when eight years old, and to have been treated at the hospital for the same condition eleven years before his death. *See Insp.* 1862, No. 61.

1608 Acute Interstitial Nephritis.

A kidney which is enlarged and weighed about fourteen ounces. Its surface is smooth, and in the recent state the cortex was uniformly white, and the veins were filled with ante mortem thrombus. Histologically the intertubular connective tissue of the organ is seen to be infiltrated throughout with small round cells.

William B., æt. 32, was admitted under Dr. Rees for pyrexia and delirium, thought to be due to typhus fever. The urine was highly albuminous, but there was no anasarca. He died in convulsions one month after admission, and at the autopsy purulent meningitis was found. There was a bed-sore upon the left buttock with thrombus in the common iliac vein. *See Insp.* 1860, No. 51.

1609   Chronic Interstitial Nephritis.

A pair of kidneys, which in the recent state weighed two and three quarter ounces. The capsules, which were adherent, have been removed, and the surface of the organs is seen to be smooth except for persistent fœtal lobulation. Histological examination shews a uniform infiltration of the kidney by a richly nucleated fibrous tissue leading to compression and atrophy of the tubules, whilst the glomeruli have thickened capsules and have undergone hyaline degeneration.

Richard G., æt. 48, was admitted under Dr. Barlow for epileptiform fits, having suffered from albuminuria for three years. He had had syphilis and was of very intemperate habits, being accustomed to drink large quantities of raw spirits. He died comatose three months after admission, and at the autopsy the remains of a small cerebral hæmorrhage were found ; the heart was of normal size. *See Insp.* 1856, No. 105.

1610   Interstitial Nephritis.

The half of a somewhat contracted kidney, the vessels of which have been partially injected. Its surface is smooth and marked by several small cysts. The cortex is diminished, and histologically shews many wasted glomeruli and dilated tubes surrounded by an excessive amount of fibrous tissue.

Caroline J., æt. 40, was admitted under Dr. Wilks for a chronic cough with œdema of the legs which had been noticed for three months. The urine was found to contain albumen, and to have a specific gravity of 1008. She died fourteen days after admission, and at the autopsy the apices of both lungs contained tubercles, and there was lardaceous disease of the spleen. The kidneys weighed 10 ounces. *See Insp.* 1873, No. 247.

1611   Granular Kidney.

A kidney which has been injected, and from which the capsule has been partially stripped, shewing the surface to be uniformly and finely granular. The organ is of about the normal size, and the fœtal lobulation is persistent. Histologically the interstitial tissue is seen to be excessive in amount, and the glomeruli are wasted.

Mary B., æt. 34, was admitted under Dr. Bright in 1831 for albuminuria of five years' duration. She died in a fit, and at the autopsy the heart was found to be of moderate size and to be healthy, with the exception of slight thickening of the mitral valve. *See Insp.* vol. 2, p. 59; and *Guy's Hosp. Reps.* 1836, p. 358.

1612   Contracted Granular Kidney.

A small kidney, which with its fellow weighed in the recent state four and a half ounces. The capsule has been stripped off, and the surface of the organ is seen to be finely granular. The cortex is wasted and contains numerous minute cysts, some superficial and others more deeply placed. Histologically the fibrous tissue is excessive in amount, the vessels thickened, and the glomeruli wasted.

Stephen A., æt. 72, was admitted under Mr. Birkett with his right humerus and femur fractured, having been run over by a cab. He died two days after admission, and at the autopsy the rest of the viscera were found to be healthy, the heart weighing 11¼ ounces. *See Insp.* 1870, No. 199.

1613   Contracted Granular Kidney.

A kidney of small size, the surface of which is in parts coarsely granular. The renal artery is atheromatous. Histologically the small vessels are thickened, the glomeruli are wasted, and the connective tissue is excessive in amount. Below is mounted the heart laid open to shew considerable thickening of the wall of the left ventricle.

Rhoda H., æt. 35, was admitted under Dr. Babington in 1847 for cerebral hæmorrhage, from which an hour later she died. At the autopsy the bladder was found to contain albuminous urine. *See Insp.* vol. 35, p. 81.

1614   Contracted Granular Kidneys.

The halves of two kidneys, which together in the recent state weighed three and a half ounces. The capsule has been removed, and the cortex is seen to be nodulated and granular. There are several small cysts. Histologically the appearances are those of advanced interstitial nephritis.

From a man, æt. 25, who died from anæmia and persistent vomiting, the state of his kidneys not having been suspected during life.

Presented by Dr. Gull.

1615   Contracted Granular Kidney.

A pair of kidneys weighing in the recent state one ounce and a half. The organs have a roughened granular surface, to which the capsule was unduly adherent. Histologically there is an excess of fibrous tissue between the tubules, some of which are larger, some smaller than normal. The glomeruli are for the most part fibroid.

Mary E., æt. 50, was admitted under Dr. Rees with anasarca, bronchitis, and albuminuria. She died in a semi-comatose condition about three months later, and at the autopsy the right pleural cavity contained 5 pints of serous fluid, and the heart was of normal weight. The renal vessels were thickened. *See Insp.* 1857, No. 199.

1616   Scarlatinal Nephritis.

The halves of two kidneys from a case of scarlet fever. The organs weighed six ounces and in the recent state presented a " mottled " appearance. The capsule has been removed and shews a smooth surface with persistent fœtal lobulation. Histologically there is acute tubal nephritis.

Laura S., æt. 6, was admitted under Dr. Gull for scarlet fever, and died three weeks later. At the autopsy there was extensive suppuration in the cellular tissue of the neck, and the pleural and peritoneal cavities contained serum and flakes of lymph. *See Insp.* 1854, No. 250.

1617 Scarlatinal Nephritis.

The kidneys of a child greatly enlarged so as to weigh eighteen ounces. They present a smooth surface marked by numerous lacerations produced post mortem. In the recent state the cortex was pale and less resistant than is normal. Histologically the kidney presents the appearances of acute glomerular, tubal, and interstitial nephritis.

Maria G., æt. 10, was admitted under Dr. Hughes about ten days after the onset of scarlet fever. She died three weeks after admission, having for the greater part of the time persistently refused to take food. At the autopsy the legs were found to be œdematous, and there was inflammatory effusion into the pleura and pericardium. The urine in the bladder was coagulable by heat. *See Insp.* 1854, No. 213.

1618 Chronic Interstitial Nephritis after Scarlet Fever.

The half of a kidney from which the capsule has been stripped, to shew a granular surface and a few superficial cysts. The organ is of the normal size, and histologically presents the appearances of chronic interstitial nephritis.

Charlotte R., æt. 36, was admitted under Dr. Washbourn with symptoms of chronic Bright's disease. At the age of 10 she had suffered from scarlatina followed by dropsy, the illness having lasted for ten months. On admission the urine was found to have a specific gravity of 1010, and to contain ·4 per cent. of albumen. Four days after admission she suddenly became unconscious and died. At the autopsy the uterus was found to contain twins 5 months old, and the heart weighed 13 ounces. *See Insp.* 1895, No. 284.

1619 Septic Nephritis.

A pair of kidneys which in the recent state shewed " minute centres of acute suppurative nephritis " in the cortex, and histologically present evidence of small-celled infiltration in the pyramids. There were numerous hæmorrhages, chiefly on the surface.

Alexander B., æt. 3 weeks, was brought to the hospital with a brawny œdema affecting the lower half of the body, and said to have been present for about three days. He died in the surgery, and at the autopsy the lower urinary passages were found to be healthy, and it was thought that the œdema was due to erysipelas. There was lymph upon the spleen. *See Insp.* 1894, No. 66.

1620   Pyæmic Abscesses of the Kidney.

The halves of two kidneys upon the surface of which are seen numerous small convex elevations, most of which are about the size of a millet seed, while some are rather larger. They are white in colour, and in the recent state exuded when incised a small bead of pus.

Henry B., æt. 16, was admitted under Mr. Hilton for an abscess at the lower end of the right femur following an injury. Pyæmia supervened, and the patient died twelve days after the onset of acute symptoms. At the autopsy septic emboli were found in the lungs, and there was inflammation of the pleura and pericardium. Abscesses were found in the heart, liver, and spleen. *See Insp.* 1855, No. 60.

1621   Ascending Nephritis. Abscesses in the Kidney.

A kidney laid open to shew within its substance numerous small abscesses, the largest of which is about one-eighth of an inch in diameter. Similar inflammatory foci project beneath the capsule of the organ and give a nodular appearance to its surface. Histologically the abscesses are associated with a diffuse infiltration of the parenchyma with small round cells.

James W., æt. 78, was admitted under Mr. Aston Key in 1835 for stone in the bladder, and died three weeks later from bronchitis. At the autopsy the mucous membrane of the pelves and ureters was reddened. The bladder was sacculated and hypertrophied, and contained puriform fluid and three calculi. *See Insp.* vol. 13, p. 35.

1622   Ascending Nephritis. Multiple Abscesses of the Kidney.

A kidney, the surface of which exhibits numerous abscesses about the size of millet seeds. The capsule

has been removed, and the surface of the organ presents
in various parts a nodular appearance from the projection of somewhat larger abscesses. The pelvis and
calices are dilated.

Elizabeth H., æt. 40, was admitted in 1832 for malignant
disease of the uterus from which she died. At the autopsy the
pelvic tumour was found to have compressed the ureters. *See
Insp.* vol. 11, p. 167.

1623   Lardaceous Kidney.

The half of a kidney, which is enlarged and on section
is seen to have a translucent appearance. The capsule
of the organ has been removed, and shews the surface
to be very slightly granular. Fœtal lobulation is
persistent. Histologically the organ is seen to be
lardaceous.

Caroline J., æt. 26, was admitted under Dr. Addison for
general anasarca with scanty albuminous urine. She had suffered
for six years from a chronic ulcer of the leg. Her ascites increased
and paracentesis was performed. The patient died a fortnight
after admission, and at the autopsy the liver and spleen were
found to be affected by lardaceous disease, the former weighing
7½ pounds. The kidneys weighed 17 ounces. *See Insp.* 1856,
No. 124.

1624   Lardaceous Kidney.   Venous Thrombosis.

A kidney which with its fellow weighed twenty-four
ounces. The cut surface shews many of the veins filled
with adherent thrombus. In the recent state the cortex
was speckled and the pyramids deeply congested.
Histologically the glomeruli and smaller vessels are
lardaceous, and there is a considerable excess of small
round cells in the connective tissue of the organ.

George B., æt. 56, was admitted under Dr. Wilks with
ulceration of the fauces, anæmia, and profuse diarrhœa. He
had previously suffered from syphilis. He died five days after
admission, and at the autopsy the intestines and thyroid gland
were found to be lardaceous. The liver and spleen appeared
normal. The left ventricle was somewhat hypertrophied, and one
of the testes was fibroid. *See Insp.* 1871, No. 267.

1625   Lardaceous Nephritis.

The half of a kidney enlarged so as to measure six and a half inches in length, the surface of which is somewhat granular. The kidney is mounted in a solution of iodine to shew the mahogany-brown staining of the lardaceous portions.

Frederick K., æt. 24, was admitted under Dr. Rees for a sinus, the result of hip disease, with which he had been affected since infancy. He died about three weeks after admission, having for the last few months of his life suffered from dropsy with albuminuria. At the autopsy the liver and spleen appeared to be free from lardaceous disease. *See Insp.* 1863, No. 172.

1626   Lardaceous Kidney.   Interstitial Nephritis.

A portion of a kidney, the surface of which is irregular and pitted. In the recent state the cut section was streaked and spotted, and the Malpighian bodies and arteries stained deeply with iodine.

Mary C., æt. 42, was admitted under Mr. Poland for pelvic cellulitis following parturition, which resulted in an abscess opening in the neighbourhood of the left hip-joint. At the autopsy the liver and spleen were lardaceous. The kidneys were of very great size. *See Insp.* 1862, No. 226.

1627   Gumma of the Kidney.

The half of a kidney, the lower end of which is infiltrated with a deposit of fine yellowish material, which under the microscope presents the structure of a syphilitic gumma. On the reverse of the specimen the capsule is seen to be thickened and to be closely adherent over the affected area.

Emma K., æt. 28, was admitted under Dr. Habershon for ascites, anasarca, and albuminuria. She died three months after admission, and at the autopsy the kidneys and spleen were found to be lardaceous. There was an abscess in the head of the pancreas, thought to be due to a softening gumma. *See Insp.* 1867, No. 234; *Guy's Hosp. Reps.* 1867, p. 391; and *Prep.* 1465.

1628   Miliary Tuberculosis of the Kidney.

A kidney stripped of its capsule and shewing on its surface numerous slightly projecting nodules, measuring about a line in diameter. Histologically they consist of a caseous centre containing giant-cells and a peripheral zone of small cells. On the reverse of the specimen is seen near the lower end of the kidney a calcareous mass embedded in the renal tissue.

Thomas F., an adult, was admitted under Mr. Bryant for urethral stricture and perineal fistulæ. Cystotomy was performed and some weeks later he died. At the autopsy a caseous mass was found in the brain, and there were miliary tubercles in the lung, peritoneum, and liver. *See Insp.* 1874, No. 197.

1629   Miliary Tuberculosis of the Kidney.

The half of a lobulated kidney, the capsule of which has been removed to shew upon the surface of the organ several white well-defined spots about one sixteenth of an inch in diameter, produced by the presence of miliary tubercles.

George B., æt. 6, was admitted under Dr. Rees for emaciation and bronchitis following an attack of measles eight weeks before admission. At the autopsy the bronchial glands were found to be caseous, and there were miliary tubercles in the lungs, liver, and spleen. *See Insp.* 1856, No. 132; and *Preps.* 263 & 1515.

1630   Miliary Tuberculosis of the Kidney.

The half of a kidney, on the surface of which are seen several small white nodules half embedded in its substance, each about the size of a millet seed. Histologically the nodules have the structure of tubercle.

George S., æt. 10, was admitted under Dr. Addison for a cough of seven weeks' duration and died three weeks later from general tuberculosis. At the autopsy the bronchial glands, lungs, liver, and meninges of the brain, pericardium, and spleen were found to be affected. *See Insp.* 1858, No. 2; and *Preps.* 1445 (50), 2007 (50) [2nd Edit.], and 1320.

1631   Scrofulous Kidney.

A kidney enlarged so as to measure six inches in length. It has been laid open to shew its lower part occupied by a large spherical cavity, lined with soft caseous material. In the upper part there are similar smaller cavities, and the persistent renal tissue is infiltrated with softening tubercle. The capsule is thickened, and where it has been stripped the surface of the organ is speckled and somewhat granular.

From a girl aged 7, who had suffered for some months from a painful abdominal tumour and died with symptoms of meningitis. See *Trans. Path. Soc.* 1875, p. 129.

Presented by Mr. Lucas.

1632   Scrofulous Kidney.   Genito-urinary Tuber-culosis.

A kidney laid open to shew the lining membrane of the pelvis and calices converted into a granular caseous material, and in parts eroded by ulceration. There are numerous cavities in the organ, the substance of which is very extensively infiltrated with tuberculous deposit. Below are mounted the ureter, bladder, and vesiculæ seminales, all of which exhibit a similar infiltration.

John S., æt. 50, was admitted under Dr. Barlow with symptoms of phthisis and a strumous abscess of a rib. He died a week after admission, and at the autopsy tuberculous disease was found in the lungs, testes, and intestine. *See Insp.* 1854, No. 192.

1633   Scrofulous Kidney.

A kidney greatly enlarged, so as to measure seven and a half inches in length. Its interior is occupied by a mass of caseous material, in which are seen numerous ragged cavities occupying the situation of the pyramids of the organ. The capsule of the kidney is somewhat thickened. Histologically the renal tissue is seen to be replaced by caseating tubercle.

1634   Scrofulous Kidney.

A kidney somewhat enlarged and incised to shew its calices dilated and lined by a rough caseous material. In the cortical portion of the kidney are numerous nodules having a similar caseous structure. On the reverse of the specimen the ureter is seen to be thickened, its mucous membrane being replaced by tuberculous deposit. Sections cut and examined ninety years after the death of the patient shewed under the microscope a characteristic tuberculous infiltration, with caseation and numerous giant-cells.

Anne B., æt. 52, was admitted under Dr. Marcet in 1807 for diarrhœa and painful micturition. She had been ill for twelve months, and died exhausted and despondent twenty-eight days after admission. At the autopsy ulcers were found in the colon, and the left kidney contained about two ounces of pus. The mucous coat of the bladder was almost entirely destroyed by ulceration. *See Old Museum Book,* No. 75.

1635   Scrofulous Kidney.

A left kidney, which is considerably enlarged, and has been laid open to shew a granular and ulcerated condition of the lining of its pelvis and calices. The organ presents several ragged excavations, the walls of which consist of a thick layer of soft yellow material, exhibiting histologically a fibro-caseous structure. On the reverse of the specimen numerous nodules are seen projecting from the surface of the organ, and the ureter is thickened and narrowed by caseous deposit in its wall. Below is mounted the bladder, the mucous membrane of which presents many rounded plaques superficially ulcerated.

1636   Tuberculous Abscess of the Kidney; Nephrectomy.

A kidney removed by operation. About the middle of the convex border is a ragged opening leading through the cortex into a cavity occupying the position of a pyramid. The cavity is lined with a granular layer of inflammatory material, in which under the microscope giant-cells and areas of caseation are recognized. The mucous membrane of the infundibula and pelvis is roughened and partially destroyed-by ulceration, and the upper part of the ureter is much thickened. At the upper end of the kidney the cortex presents numerous nodular prominences resulting from tuberculous infiltration.

William F., æt. 21, was admitted under Mr. Lucas in 1894 with symptoms suggesting stone in the bladder. Median cystotomy was performed, but no calculus was found. Subsequently the kidney was explored and an abscess in its cortex incised. A month later the kidney was removed, and the patient, who made a rapid recovery, reported himself as in good health ten months after the operation. *See Surgical Reports*, vol. 172, Case 170.

1637   Tuberculous Ulceration of Pelvis and Ureter.

A kidney with its pelvis and ureter laid open to shew an ulcer encircling the former for a distance of an inch, its lower limit corresponding to the commencement of the ureter. The border of the ulcer is sinuous and well defined, and its base is formed of granular material, which histologically shews evidence of caseation. There is a small superficial ulcer in the ureter. On the reverse of the specimen miliary tubercles are seen upon the outer surface of the pelvis.

1638   Tuberculous Pyonephrosis.   Nephrotomy.

The posterior half of a left kidney partially detached from an investment of dense fibrous tissue. The cut surface of the organ shews the pelvis and calices to be dilated and lined by a thick shaggy layer of fibrocaseous material. The exterior of the kidney is irregularly nodulated, and presents numerous minute cysts. A blue rod indicates the opening left after lumbar nephrotomy.

Edward E., æt. 32, was admitted under Dr. Moxon with a tumour in the left loin associated with pyuria. He had suffered from symptoms of renal disease for one year. The tumour was explored by Mr. Bryant and ten ounces of pus were evacuated. At the autopsy the right kidney was healthy and shewed compensatory hypertrophy. The spleen was firmly adherent to the left kidney. *See Insp.* 1870, No. 175.

1639   Tuberculosis of the Ureter.

The lower portion of the ureters with the bladder laid open to shew beneath the mucous surface of the right ureter numerous somewhat flattened nodules, varying in size from a millet seed to a split pea. The last inch of the ureter is considerably thickened, and the bladder presents patches of superficial ulceration. Histological examination shews that the nodules in the ureter have the structure of tubercle.

William W., æt. 45, was admitted under Dr. Perry for dyspnœa and anasarca associated with hæmaturia. Ten days later he died in a state of coma, and at the autopsy both kidneys were affected by tubal and interstitial nephritis, and at the lower end of the right kidney were several tuberculous excavations. The heart weighed 17 ounces. *See Insp.* 1889, No. 91.

1640   Cretaceous Kidney.

The half of a kidney in a condition resembling hydro-nephrosis. One of its cavities is seen to be filled with a chalky substance, and in the recent state a similar material occupied the other dilated spaces. Histological examination shews that the persistent tissue is infiltrated with caseous deposit.

Thomas H, æt. 30, was admitted under Mr. Bryant for disease of the left hip-joint, and died three weeks later from perforative ulceration of the colon. At the autopsy a stricture of the urethra was discovered. There was a small abscess in one of the pyramids of the other kidney. *See Insp.* 1870, No. 314; and *Prep.* 831.

1641   Cretaceous Kidney.

A left kidney of normal size and shape, which has been laid open to shew extreme atrophy of its secreting tissue from dilatation of its pelvis and calices. These spaces are filled with a white substance, which in the recent state resembled " prepared chalk, with just a sufficient quantity of water to give it plasticity."

William T., æt. 60, was admitted under Mr. Bransby Cooper for stricture of the urethra and a urinary fistula opening at the umbilicus. Attempts were made to close the fistula, and the patient died from purulent peritonitis. At the autopsy the right kidney and its ureter were dilated and contained puriform fluid. *See Insp.* vol. 11, p. 7.

1642   Cretaceous Kidney.

A kidney of small size, the dilated pelvis and calices of which are filled with a mortar-like material. One of the cavities at the upper end shews a partially detached lining consisting of a shell of calcareous deposit.

Mary A., æt. 46, was admitted under Mr. Durham for injuries from which two days later she died. At the autopsy the right ureter was found to be impervious at its upper end and filled with cretaceous matter below. The left kidney was hypertrophied and contained numerous cysts. *See Insp.* 1892, No. 147.

1643   Squamous-celled Epithelioma of the Kidney.

A kidney slightly enlarged and retaining its normal shape. It has been laid open to. shew its substance infiltrated by a hard white growth, having the histological characters of a squamous-celled epithelioma. Beneath the capsule of the organ are numerous small cysts.

From a patient whose leg was amputated by Mr. Aston Key for cancer. At the autopsy secondary deposits were found in the skin and heart-muscle. *See Preps.* 1248 (80), 1390, 1641, and 1658 (2nd Edit.).

1644   Spheroidal-celled Carcinoma of the Kidney.

A kidney laid open to shew at its upper end a circum-scribed globular tumour, measuring two and a half inches in diameter. The growth has a thick fibrous capsule, and its cut surface presents a brown hue from extravasated blood. It is traversed by several fibrous trabeculæ. Histologically the growth is a spheroidal-celled carcinoma with abundant stroma.

1645   Spheroidal-celled Carcinoma of the Kidney.

A kidney measuring six inches in length and ex-tensively infiltrated with nodular masses of growth, some brown and firm, and others undergoing degenerative softening. On the reverse of the specimen the surface of the kidney is seen to be marked by irregular depres-sions. Histologically the growth is a spheroidal-ceiled carcinoma with scanty stroma.

Emma T., æt. 26, was admitted under Mr. Howse with an abdominal tumour in the region of the right kidney, associated with hæmaturia and emaciation. During the operation of laparo-tomy the patient suddenly died, and at the autopsy it was found that a portion of the growth which had invaded the renal vein had become detached and produced pu'monary embolism. There were secondary deposits in the lungs and liver and in the lumbar lymphatic glands. *See Insp* 1886, No. 82.

1646   Adrenal Tumour of the Kidney.

A kidney partly embedded in which and projecting from its convex border is a nodular mass of growth measur-ing two inches in diameter. The growth is seen to extend as far as the pelvis of the organ, and to be partially encap-suled. Histological examination shews it to possess the characters of the renal tumours which spring from suprarenal tissue embedded in the cortex of the kidney. *See Trans. Path. Soc.* 1896, p. 123.

1647   Spheroidal-celled Carcinoma of the Kidney.

A right kidney of normal size, the structure of which
is for the most part replaced by a mass of growth which
in the recent state was soft and yellowish, "something
like decolourized fibrin of blood." A section has been
made through the kidney, and shews that the growth
consists of an aggregation of nodules partially separated
from each other by fibrous tissue. The pelvis is dilated,
and at either end of the organ there is a shell of persistent cortical substance. On the reverse of the specimen
a mass of growth is seen occupying the hilum and
projecting into the renal vein. Histologically the growth
is a medullary spheroidal-celled carcinoma with areas
of colloid degeneration.

Margaret H., æt. 57, was admitted under Dr. Pavy for chronic
bronchitis and emphysema, from which five days later she died. At
the autopsy the left kidney was small and granular. No malignant
growth was found in any other part of the body. See Insp. 1873,
No. 219; and Trans. Path. Soc. 1876, p. 204.

**1648   Spheroidal-celled Carcinoma of the Kidney.
Plugged vein and Ureter.**

A kidney enlarged so as to measure seven inches in
length and uniformly infiltrated by a soft new growth
which projects into the dilated pelvis of the organ and
blocks the renal vein. The upper end of the ureter is
seen to be distended for a distance of about two inches
by a mass of growth continuous with that in the pelvis.
The growth has the histological characters of a spheroidal-
celled carcinoma with scanty stroma.

1649   Cylindrical-celled Carcinoma of the Kidney.

A left kidney somewhat enlarged and laid open to shew
its substance thickly beset with soft nodules of growth
measuring an eighth of an inch to an inch in diameter.
On the reverse of the specimen similar nodules are seen
to project from the surface of the organ, and there

is one prominent bossy growth measuring two and a half inches across, to which the capsule is adherent. Histologically the deposit consists of a fibrous stroma, the alveoli in which are lined with cubical epithelium, some of them being also filled with spheroidal cells.

Sarah W., æt. 58, was admitted under Dr. Hughes for hæmaturia of three months' duration. She had been ailing for about a year, and for two months had been unable to walk. After admission the paraplegia increased and bed-sores appeared. She died about seven weeks later, and at the autopsy a mass of malignant growth was found involving the vertebræ and extending into the spinal canal. There were secondary deposits in the lungs and in the lumbar and bronchial glands. *See Insp.* 1858, No. 11.

1650   **Cylindrical-celled Carcinoma of the Kidney.**

A left kidney, its upper two-thirds occupied by a mass of soft vascular growth. The deposit has a trabeculated structure and histologically consists of cylindrical-celled carcinoma, many alveoli containing extravasated blood. The growth has distended the capsule, and presents several nodular excrescences.

John C., æt. 55, was admitted under Dr. Bryant for a fracture of the skull, from the effects of which he died. At the autopsy the left kidney was found to weigh 23 ounces, and parts of the growth were noticed to have undergone colloid degeneration. *See Insp.* 1875, No. 494.

1651   **Carcinoma of the Kidney. Hydronephrosis.**

The half of a kidney enlarged and deformed by a deposit of new growth. Its measures seven inches in length and four and a half inches transversely. The pelvis and calices are dilated, and the interior is roughened by invasion of the growth. The renal substance is entirely replaced by a mass of soft white material, which has the histological characters of a spheroidal-celled carcinoma with scanty stroma.

1652   **Villous Carcinoma of the Kidney. Hydrone-phrosis.**

Portions of a kidney enlarged so as to measure seven

inches in length, the pelvis and calices of which are dilated and contain several calculi.   To the lining membrane of certain of the dilated spaces is attached a growth of villous type, some of the processes being short and warty, whilst others are filamentous or polypoid.   The capsule of the organ is greatly thickened, and at one point there is a fistulous communication between the interior of the kidney and the adherent colon.   Histologically the growth has the structure of a cylindrical-celled carcinoma.   *Presented by Dr. Bright.*

1653   Kidney invaded by Carcinoma.

The half of a kidney in the hilum of which is a large mass of growth invading and destroying the upper half of the organ.   The cut surface of the growth shews it in parts to have the gelatinous appearance of colloid degeneration.   Histologically it is a spheroidal-celled carcinoma.

From Mrs. C., æt. about 50, who three years before her death had her breast removed for malignant disease.   Subsequently secondary deposits occurred in various parts of her body, and at the autopsy growth was found in the skull and brain, and in the lumbar lymphatic glands.   *See Insp.* vol. 10, p. 106 ; and *Preps.* 1585 (75), 1004 (50), and 2275 (40), 2302 (80) [2nd Edit.].

Presented by Dr. Hodgkin, 1831.

1654   Carcinoma at the Hilum of the Kidney.

A right kidney, at the inner margin of which is situated an oval tumour measuring three and a half inches in its longest diameter.   It is closely applied to the hilum of the kidney, and the renal vein is seen crossing its anterior surface.   The kidney has been laterally compressed, and moulded to the convexity of the tumour.   Anteriorly the growth is nodular, some of the nodules projecting into the dilated renal vein.   On the back the surface is rough, and was firmly adherent.   The cut section shews the growth to be composed of homogeneous material traversed by coarse strands of translucent fibrous tissue. Histologically it has the structure of a spheroidal-celled carcinoma with abundant stroma.

Ann W., æt. 29, was admitted under Dr. Shaw for a painful tumour in the region of the right kidney which had been noticed to be slowly increasing in size for eighteen months. Laparatomy was performed, and the tumour together with the right kidney was removed. The patient died a few hours after the operation. At the autopsy the vena cava was found to have been lacerated, the opening being securely ligatured. One of the aortic lymphatic glands was enlarged by secondary deposit. *See Insp.* 1891, No. 220.

1655   **Sarcoma of the Kidney.   Hydronephrosis.**

A kidney enlarged so as to measure five inches in length, and uniformly infiltrated by growth which in parts is soft and nodular, whilst elsewhere it is firm and fibrous-looking. The calices are dilated, and on the reverse of the specimen a considerable mass occupies the hilum of the organ, and has obstructed the ureter. Histologically the growth is a round-celled sarcoma.

From a child whose liver contained small scattered nodules of a similar deposit.

Presented by Mr. Pearse.

1656   **Round-celled Sarcoma of the Kidney.**

The two halves of a right kidney which is uniformly enlarged, measuring five and a half inches in length. Its cut surface shews it to be universally infiltrated with a soft white homogeneous material, which histologically has the characters of a small round-celled sarcoma.

George S., æt. 37, was admitted under Mr. Bryant for an enormous periosteal sarcoma of the left femur. He died from acute peritonitis three months after admission, and at the autopsy secondary deposits were found in the intestine, dura mater, lumbar lymphatic glands, and left kidney. *See Insp.* 1857, No. 156.

1657   **Sarcoma of the Kidney.**

The half of a kidney, the surface of which presents a

hobnailed appearance from irregular infiltration with new growth, which in the recent state was very vascular. Histologically the deposit has the characters of a small round-celled sarcoma.

> Henry B., æt. 9, was admitted under Mr. Lucas for a growth in the superior maxilla, having some years before been operated upon for cleft palate. The growth sloughed, and the patient died a week after his admission. At the autopsy the kidneys were found to be more than twice their normal weight, and there were secondary deposits in several other viscera. *See Insp.* 1892, No. 103.

1658   Sarcoma of the Kidney.

A left kidney, the lower half of which is occupied by a mass of growth nearly as large as the normal organ. Its surface is nodulated, and its cut section shews it to consist of masses of growth, some encapsuled and some infiltrating the renal tissue. Histologically it has the characters of an oval-celled sarcoma, some portions being highly vascular and others necrotic.

> Thomas T., æt. 64, was admitted under Dr. Goodhart for pain in the chest and emaciation of four months' duration, associated with a small soft growth on the sternum. He died two months after admission, and at the autopsy secondary deposits were found in the lungs, liver, and abdominal lymphatic glands. *See Insp.* 1895, No. 95.

1659   Sarcoma of the Kidney.   Hydronephrosis.

A kidney affected by hydronephrosis, and enlarged so as to measure ten inches in length. A section has been made through it, and shews that much of the enlargement is due to infiltration of the renal tissue by a growth which is in some parts soft and yellow, whilst in others it is white, firm, and translucent. Histologically the deposit has the structure of a spindle-celled sarcoma.

> From Mr. E. E. L., æt. 53, who had suffered for ten years from renal pain and hæmaturia followed by the appearance of a

tumour situated in the region of the kidney. At the autopsy no secondary deposits were found in any organ. *See Trans. Path. Soc.* 1869.

Presented by Dr. Wilks.

1660   Sarcoma of the Kidney. Nephrectomy.

A left kidney, the lower half of which is occupied by a globular mass of soft growth measuring three and a half inches in diameter. At the upper end of the tumour a nodule is seen projecting into the dilated pelvis, and obstructing the orifice of the ureter. The growth is limited by the capsule of the organ, the surface of which presents persistent fœtal lobulation above, while below, over the growth, it is somewhat nodular. Histologically the tumour is composed of round and fusiform cells, some of the latter presenting ill-marked transverse striation.

Winifred J., æt. 14, was admitted into the Waterloo Road Hospital, under Mr. Jacobson, for a tumour in the region of the left kidney, which was removed through the loin. The patient made a good recovery from the operation, but died about four months later from secondary deposits. *See MS. Notes,* p. 15.

1661   Melanotic Sarcoma of the Kidney.

A portion of a kidney, projecting from the surface of which is a small rounded mass of growth measuring a third of an inch in diameter. The cut surface shews it to be triangular on section, resembling an infarct. Histologically the growth is a sarcoma consisting of oval cells, many of which contain brown pigment.

Ann O., æt. 40, was admitted under Dr. Moxon for ascites. A month later she died, and at the autopsy melanotic growth was found in the skin, pleura, peritoneum, intestine, lungs, liver, and spleen. *See Insp.* 1885, No. 90; and *Preps.* 933 & 1526.

1662   Melanotic Sarcoma in the Perinephric Tissue.

The half of a kidney surrounded by the perinephric fat. Embedded in the fat are seen numerous rounded masses

of black growth varying from an eighth of an inch to
an inch in diameter. A similar nodule an inch across
occupies the hilum of the organ. The kidney appears
to be free from deposit. Histologically the growth is
a round-celled melanotic sarcoma. *See Preps.* 1551,
1555, and 1661 [2nd Edit.].

Presented by Sir Astley Cooper.

1663   Kidney invaded by Sarcoma.

A kidney, at the hilum of which is seen a large mass of
dark growth which has compressed the ureter and produced a slight dilatation of the pelvis of the organ.
Histologically the growth has the structure of a sarcoma
with round and oval cells.

1664   Fœtal Cystic Kidney.

A kidney measuring an inch and a half in length,
honeycombed throughout by small cysts. Histologically
no trace of renal structure is discovered. The cysts
have thin walls, and are lined with flattened cubical
epithelium.

From a seven months male fœtus which was stillborn.

Presented by Mr. Harold Hodgson, 1891.

1665   Fœtal Cystic Kidneys.

The kidneys of an anencephalous fœtus, one of which
appears to be normal, whilst the other is for the most
part converted into a congeries of thin-walled cysts.
The cysts are largest and most numerous in the cortical
portion. The pelvis and upper part of the ureter are
dilated, and it was remarked that in the recent state the
ureter was impervious.

1666   Fœtal Cystic Kidney.

A portion of a kidney which with its fellow weighed in
the recent state two pounds and three quarters. The

cut surface shews the organ to be thickly beset with minute cysts uniform in size, and none of them larger than a millet seed. Histologically they are seen to be lined by cubical epithelium, and some of them to be filled with proliferated cells.

From a fœtus which could not be delivered until after perforation of the thorax had been performed.

Presented by Dr. J. S. Johnson, 1877.

1667   Cystic Kidneys.   Renal Calculus.

Two kidneys of about equal size and measuring six inches in length. Both are cystic, the pelvis and ureter of the left being normal, whilst the pelvis of the right is slightly dilated and presents a rough dark brown calculus impacted at the commencement of the ureter. Histologically some of the cysts are seen to be surrounded by fairly normal renal parenchyma with dilated tubules.

1668   Cystic Kidney.

A kidney, from the lower end of which projects a thinwalled globular sac measuring six inches in diameter, which in the recent state was filled with pultaceous material. The kidney itself is considerably enlarged, and its surface is covered with rounded prominences varying in size from a pea to a pullet's egg, which on section are seen to be cysts, some of them containing a colloid substance.

Ann O., æt. 81, was admitted under Mr. Cock with a strangulated hernia, for the relief of which herniotomy was performed. She died shortly after the operation, and at the autopsy a perforation was found in the small intestine, and there was acute general peritonitis. The other kidney appears to have been normal. *See Insp.* 1859, No. 69.

1669   Cystic Kidney.

A left kidney measuring eight inches in length and three inches transversely. It has been laid open to shew its substance converted throughout into cysts, some of which are no larger than a millet seed, whilst others measure more than an inch in diameter. Histologically the cysts are bounded by fibrous tissue lined with flattened epithelium.

Caroline P., æt. 41, was admitted under Dr. Habershon with albuminuria and died comatose on the following day. At the autopsy the lungs were covered with recent lymph, and there was œdema of the larynx. Both kidneys were similarly affected, and weighed, the right 84, and the left 53 ounces. The heart was somewhat hypertrophied. *See Insp.* 1867, No. 21.

1670   Cystic Kidney.

The half of a cystic kidney measuring about seven inches in length. The pelvis and ureter are of normal size. Histologically the renal tissue is fibrous with wasted glomeruli. The cysts are lined with cubical epithelium.

From a patient whose liver presented a similar cystic condition. *See Prep.* 1372, and *Trans. Path. Soc.* vol. 7, p. 235.

Presented by Mr. Key.

1671   Cystic Kidney.

The half of a kidney the cortical portion of which is occupied by numerous cysts varying from a twelfth to a third of an inch in diameter, and filled with gelatinous contents. The kidney is little if at all enlarged, and except as regards its cortex appears histologically to be normal.

1672   Cystic Kidney.   Nephrectomy.

A kidney weighing in the recent state 20 ounces, laid open to shew the secreting structure almost entirely replaced by cysts, separated from each other by thin membranous septa. At the upper end of the organ is a small area of renal tissue free from cysts, but presenting dilated calices.

The kidney was removed from a boy, æt. 2½, by Mr. Jacobson in 1890.

1673   Cystic Kidney.

A left kidney somewhat enlarged, and laid open to show numerous small cysts, some of which in the recent state contained a brownish colloid substance. The cysts are for the most part situated in the pyramidal portion of the organ, very few of them being visible upon the surface. They vary in size from a line to a third of an inch in diameter. Histological examination shews them to be lined by a cubical epithelium, the surrounding renal structure being normal.

Jane II., æt. 48, was admitted under Mr. Jacobson for carcinoma of the breast, which was removed by operation. Five days later she died, and at the autopsy the right kidney was found to be slightly scarred, but otherwise healthy. *See Insp.* 1804, No. 78.

1674   Cystic Kidney ; Nephrectomy.

A left kidney, measuring six and a half inches in length and three in width. It has been divided to shew the greater part of its substance converted into a congeries of thin-walled cysts, some of which are very minute, whilst others are as much as two inches in diameter. On the reverse of the specimen the ureter is seen to be small, and at the upper end of the organ a considerable portion of cortex is unaffected by the disease.

Olive E., æt. 30, was admitted under Mr. Lucas in September 1892, having five months previously suffered from several attacks of renal colic. Since that date she had experienced occasional pain in the left lumbar region, and three months before admission a tumour had been detected in the region of the left kidney. On admission the urine was found to have a specific gravity of 1024, and to contain a normal percentage of urea. There were no abnormal constituents. The kidney was removed and the patient was discharged, well, six weeks after the operation. *See Surgical Reports,* vol. 100, Case 186.

1675   Cystic Kidney.

A left kidney greatly enlarged, so as to measure in the

recent state nine inches in length and four inches in
width.   Its surface is very irregular from the presence
of numerous projecting cysts, varying in size from a
millet seed to a pigeon's egg.   On section, similar cysts
are seen to occupy almost the entire organ.   Histologically the cavities are lined with flattened epithelium.

Frank L., æt. 40, was admitted under Dr. Shaw for dyspnœa. A swelling was noticed in the left hypochondrium, and the
urine, which had a specific gravity of 1010, contained a small
quantity of albumen and 1·2 per cent. of urea.   On the day after
admission symptoms of urœmia supervened, and on the following
day he died comatose.   At the autopsy the heart was found
to weigh fourteen and a half ounces, and the two kidneys fifty
ounces.   The right kidney was in a similar condition but was
slightly smaller.   Some of the cysts were filled with a light-
coloured, highly albuminous fluid, containing 2 per cent. of urea,
whilst in others the secretion was dark brown or black from the
presence of altered blood.   *See Insp.* 1892, No. 323.

1676   Calcified Cyst of the Kidney.   Hydronephrosis.

A right kidney laid open to shew at its lower end an
ovoid cyst measuring 4 inches in length and filled, in
the recent state, with a brownish fluid containing
cholesterine.   The wall of the cyst is thick and calcified,
and its smooth fibrous lining presents numerous lacunæ.
Above the cyst the kidney is extremely hydronephrotic,
and histological examination of its cortex shews the
dilated spaces to be bounded by a narrow layer of fibrous
tissue in which are seen renal tubes filled with colloid
contents.

Michael K., æt. 35, was admitted under Dr. Habershon for
acute alcoholic gastritis, and died in the course of a few hours.
At the autopsy the left kidney was found to weigh 8 ounces,
and its structure appeared normal.   The right ureter was adherent
to the wall of the calcified cyst, and where adherent was occluded.
See Insp. 1879, No. 412.

1677   Cyst in the Kidney.

A right kidney, partially imbedded in which and projecting from its lower end is seen an ovoid cyst

measuring about $2\frac{1}{2}$ inches in its longest diameter. The interior of the cavity is smooth and presents no communication with the pelvis. Its wall is thin and fibrous and separable from the capsule of the organ. The surface of the kidney is slightly granular, and histologically some of the glomeruli are seen to be fibroid.

1678   Cysts of the Ureter and Pelvis of the Kidney.

A kidney with its pelvis and ureter laid open to display upon their mucous surface a number of small translucent nodules measuring about a line in diameter. Histological examination shews that the nodules have a fibrous investment covered by mucous membrane and containing masses of granular material. The kidney itself is scarred and fibroid.

Charles McL., æt. 30, was admitted under Dr. Bright in 1836 for anasarca and dyspnœa. He was a house painter, and two years previously had suffered from lead colic. Four days after admission he died, and at the autopsy both pleural cavities were found full of serous fluid, and the heart was greatly enlarged. The aortic valves were diseased. *See Insp.* vol. 22, p. 98, and Drawing 364; *Prep.* 1413 (o5) [2nd Edit.]; and *Trans. Path. Soc.* 1890, p. 170.

1679   Hydatid of the Kidney.

A kidney presenting on its convex surface an ovoid cyst measuring three inches in its longest diameter. The cyst has been laid open to shew that there is no communication between its cavity and the pelvis of the kidney. The membrane and daughter cysts which it contained are at the bottom of the preparation-jar.

From a man who committed suicide by taking prussic acid.

Presented by Dr. Roper, 1864.

1680   Suppurating Hydatid of the Kidney.

A left kidney seen from behind and mounted to shew

its lower half replaced by a large globular cyst measuring
six inches in diameter. The wall of the cyst is fibrous
and in parts calcified. In the recent state it was lined
with hydatid membrane and contained numerous
daughter cysts.

Samuel D., æt. 39, was admitted under Dr. Habershon with
enlargement of the spleen. There was also a rounded elastic
swelling on the left side of the abdomen, apparently distinct from
the splenic tumour. The blood contained an excessive proportion
of white corpuscles. The patient died five weeks after admission,
and at the autopsy the wall of the cyst was found to be suppurating.
See Insp. 1875, No. 3.

1681   Hydatid of the Kidney.

A left kidney, the lower half of which is occupied by
a globular cyst, which in the recent state contained
more than two pints of clear fluid and was lined with
hydatid membrane.

Mary T., æt. 23, was admitted under Mr. Durham for a
malignant growth of the shoulder, and died from shock following
amputation. At the autopsy a second hydatid cyst was found in
the left lobe of the liver. *See Insp.* 1875, No. 414.

1682   Hydatid of the Kidney.

A left kidney, from the hilum of which protrudes a thin-
walled cyst of about the same size as the organ itself.
The pelvis and ureter, which are somewhat dilated, are
seen in front of and closely adherent to the sac. In the
recent state the cyst contained about a hundred daughter
cysts, some of which are seen at the bottom of the
preparation-jar.

James B., æt. 56, was admitted under Dr. Rees for phthisis,
from which ten weeks later he died. *See Insp.* 1857, No. 43.

1683   Hydatid of the Kidney.

A right kidney, from the upper end and convex border
of which projects a globular cyst about the size of a

tangerine orange. In the recent state the cyst was lined with the hydatid membrane and contained the daughter cysts which are seen at the bottom of the preparation-jar.

From a patient in whose liver also hydatids were found.

•

1684   Hydronephrosis.   Nephrectomy.

A left kidney measuring seven and a half inches in length, laid open to shew the pelvis, infundibula, and calices greatly dilated. The pyramids have undergone atrophy, and only a thin shell of cortical substance remains.

Alfred H., æt. 20, was admitted under Mr. Lucas in 1892 for pain in the left loin associated with pyuria. His illness began four years previously with an attack of colic during which a renal tumour appeared but quickly subsided. The patient was quite well after this until three months before admission, when the pain for which he sought treatment commenced. His urine was found to have a specific gravity of 1016 and to contain pus, blood, and crystals of triple phosphate. No tumour could be felt. The kidney when removed contained 8½ ounces of cloudy yellow fluid with a specific gravity of 1010 and yielding 1·2 per cent. of urea. The patient was discharged well 25 days after the operation. *See Surgical Reps.* vol. 160, Case 176.

1685   Hydronephrosis.   Nephrectomy.

A left kidney measuring about nine inches in length, and presenting an extreme condition of hydronephrosis. The pelvis has been cut away to shew the openings into the dilated calices. On the reverse of the specimen is seen a ragged aperture, from which the organ was drained through the loin.

Edith T., æt. 17, was admitted under Mr. Symonds in 1888 with a urinary sinus which had persisted since the incision of a hydronephrosis five months previously. The kidney was removed and the patient made a good recovery. *See Surgical Reps.* vol. 138, Case 85.

1686   Hydronephrosis. Ureter occluded by Calculus.

A right kidney converted into a multilocular sac, the wall of which is about a sixteenth of an inch in thickness, and consists of fibrous tissue in which closely packed wasted glomeruli can be detected under the microscope. The sac retains the shape of the kidney, and measures five and a half inches in length. The ureter is considerably dilated throughout, and just above the bladder it is completely occluded by a calculus.

> Ellen W., æt. 26, was admitted under Dr. Goodhart for hæmaturia and pain in the left loin, having thirteen years previously and on several subsequent occasions passed renal calculi. A tumour was felt in the left loin, from which pus was evacuated. The patient died nine days after the operation, and at the autopsy an abscess was found around the left kidney, the pelvis of which was enlarged and contained purulent urine and several calculi. *See Insp.* 1891, No. 447.

1687   Hydronephrosis.

A portion of a right kidney, with a mass of growth surrounding the aorta and invading the duodenum. The pelvis and calices of the kidney are greatly dilated, as the result of the pressure of the growth upon the ureter. Histologically the growth is a small-round-celled sarcoma.

> George C., æt. 46, was admitted under Dr. Taylor with hæmaturia. A tumour was felt on the right side of the abdomen. It increased rapidly, and the patient died five weeks after admission. At the autopsy an enormous mass of growth was found in front of the vertebræ, which invaded and completely occluded the right ureter. The growth was thought to have its primary seat in the aortic lymphatic glands. *See Insp.* 1891, No. 396.

1688   Hydronephrosis.   Hypertrophy and Dilatation of the Ureter.

A kidney with its ureter mounted to shew the effects of obstruction. The pelvis and the first three inches of the ureter are little if at all affected, but the calices are considerably dilated, and the lower part of the ureter measures an inch and three quarters in diameter at its widest part. Its walls are much hypertrophied.

1689 Hydronephrosis. Ureter terminating in the Urethra.

The right kidney of a child mounted to shew a condition of hydronephrosis associated with dilatation of the ureter. Below are mounted the bladder, urethra, and the terminal portion of the rectum. The ureter opens into the urethra at the situation of the prostate, and the rectum communicates with the same canal three quarters of an inch beyond this point. The left kidney was rudimentary. *Presented by Mr. Farrant Fry.*

1690 Hydronephrosis. Renal Calculi.

The kidneys with the ureters and bladder of a child. Both kidneys are enlarged by distension of their pelves and calices, in which are contained large white calculi.

Presented by Mr. F. Toulmin.

1691 Hydronephrosis. Renal Calculi.

A left kidney, in a condition of extreme hydronephrosis, laid open to shew the dilated pelvis and calices occupied by large branching calculi, the extremities of which are smooth and white, whilst elsewhere their surface is covered with a rough brown deposit. A yellow rod has been passed into the ureter, which is dilated and embedded in a mass of malignant growth.

Thomas L., æt. 47, was admitted under Dr. Goodhart with pus and blood in his urine and symptoms of vesical calculus. Twenty-one years previously a large stone had been removed from his bladder. He died from gangrenous bronchopneumonia about three months after his admission and at the autopsy a large

M 2

growth was found on the posterior wall of the bladder obstructing the left ureter. The right kidney weighed five ounces and was cirrhotic. There were secondary deposits in the aortic glands and in the liver. *See Insp.* 1880, No. 354.

1692   Hydronephrosis.   Renal Calculi.

A kidney, somewhat enlarged, from which portions of the cortex have been removed to display brown calculi lodged in the dilated calices. The pelvis is enlarged and filled with similar stones; and the ureter is hypertrophied and dilated.

1693   Hydronephrosis.   Ureter kinked over Renal Vein.

A left kidney seen from behind and presenting a globular dilatation of its pelvis. At its origin the ureter is acutely bent over a large branch of the renal vein entering the lower end of the organ. In the recent state the hydronephrotic sac could not be evacuated by pressure until the relation of the parts was disturbed.

Sarah N., æt. 36, was admitted under Dr. Pye-Smith for epileptiform fits which continued till her death twelve hours after admission. At the autopsy the right kidney was found to weigh seven and a half, the left three and a half ounces. *See Insp.* 1880, No. 424; and *Trans. Path. Soc.* 1804, p. 107.

1694   Hydronephrosis.   Ureter compressed by Renal Vessels.

A left kidney, seen from behind. It is enlarged so as to weigh seventeen and a half ounces, and presents a condition of hydronephrosis. The pelvis is dilated into a globular sac two and a half inches in diameter. Coursing along the inner edge of this sac to enter the lower end of the kidney is seen a branch of the renal artery with its corresponding vein. The ureter, which arises from the pelvis just below the level of these aberrant vessels, is bent over them and kinked.

Albert C., æt. 23, was admitted under Mr. Davies-Colley for a sarcoma of the lower jaw. The growth was removed, erysipelas supervened, and the patient died nine days after the operation from hæmorrhage from the lingual artery. At the autopsy the right kidney was found to weigh six ounces. *See Insp.* 1887, No. 421 ; and *Trans. Path. Soc.* 1894, p. 108.

1695   Hydronephrosis.

A pair of kidneys of which the right presents a moderate degree of hydronephrosis. Its pelvis is dilated and its ureter is acutely bent over a descending branch of the renal artery, which is seen tightly stretched across the sac. The left kidney shews compensatory hypertrophy.

Alexander G., æt. 17, was admitted under Dr. Pitt for tuberculous disease and died about six weeks later. At the autopsy the cervical and mesenteric glands were caseous and there were miliary tubercles in the lungs, spleen, kidneys, and liver. *See Insp.* 1891, No. 464 ; and *Trans. Path. Soc.* 1894, p. 109.

1696   Hydronephrosis. Calculus impacted in Ureter.

A kidney of normal size laid open to shew considerable dilatation of the pelvis and calices. The ureter is also dilated and a little above its entrance to the bladder it is occluded by a dark brown calculus.

Presented by Mr. Aston Key.

1697   Double Hydronephrosis. Vesical Calculus.

The kidneys, ureters, and bladder mounted to shew bilateral hydronephrosis and dilatation of the ureters. The kidneys are unequal in size, and the ureter of the left, which is the larger, is acutely kinked at its upper end. The bladder is of hourglass shape, the smaller sac, into which the ureters open, being in the recent state entirely filled by a large stone.

William H., æt. 31, was admitted under Mr. Cooper Forster with symptoms of vesical calculus from which he had suffered at intervals during fourteen years. Lithotrity was attempted and the patient died three weeks after his admission. At the autopsy the calculus in the bladder was found to be partly crushed and to weigh 3¾ ounces. *See Insp.* 1868, No. 231.

1698   Hydronephrosis, with Atrophy of the Kidney.

An atrophied right kidney which in the recent state weighed less than an ounce. It measures two and a half inches in length, and consists of a thin shell of cortical substance surrounding the dilated pelvis and calices. The orifice of the ureter is blocked by a stone lodged in the pelvis.

William H., æt. 34, was admitted under Dr. Pye-Smith with phthisis from which he died. At the autopsy the left kidney was found to be enlarged and hydronephrotic. Its texture was coarse and a large calculus occupied its pelvis. *See Insp.* 1879, No. 128.

1699   Hydronephrosis.   Congenital   Stenosis   of Ureter.

A kidney of a child seven years of age enlarged so as to measure six inches in length. It is converted into a membranous sac covered in parts by a thin shell of renal tissue. The pelvis is much dilated, the ureter being so narrow as barely to admit a small probe.

1700   Hydronephrosis from Prolapsus Uteri.

The kidneys with the ureters and female genital organs. The kidneys are in a condition of extreme hydronephrosis and the ureters are dilated. The uterus is prolapsed and has carried forward with it the bladder, so that in the recent state the lower ends of the ureters were compressed between the uterus and the pelvic arch.

Mary A., æt. 38, was admitted under Dr. Fagge in a moribund condition. Her urine contained pus and for about a year before admission she had suffered from prolapse of the uterus.

She died three days later, and at the autopsy her liver shewed signs of tight-lacing; "the quantity of the renal parenchyma left was not one-quarter of the whole and had quite lost its visible structure." *See Insp.* 1869, No. 148.

1701   Hydronephrotic and Cystic Kidney.

A kidney measuring seven inches in length and laid open to shew the renal tissue to be almost entirely absorbed as a result of hydronephrosis combined with cystic degeneration.   On the right side of specimen are seen the dilated calices and pelvis from which a yellow rod has been passed into the ureter, whilst on the left is seen a congeries of thin smooth-walled cysts which histologically are found to be lined with flattened epithelial cells.   *Presented by Dr. Dowler.*

1702   Hydronephrosis.   Nephrectomy.

A kidney measuring seven inches in length divided by a frontal section.   Its lower end is much enlarged, and is occupied by four thin-walled cysts which have replaced the renal tissue, the largest measuring $3\frac{1}{2}$ inches in diameter.   The upper portion of the organ is normal and a pink rod has been passed from its pelvis into the ureter.   A dark brown calculus is lodged in what appears to have been the mouth of a second ureter forming the excretory duct of the lower half of the organ.

From a woman who was admitted for pain in the back, associated with a renal tumour.   The kidney was removed and four years later the patient was in good health.

Presented by Mr. Lucas.

1703   Calculous Hydronephrosis.   Carcinoma of the Pelvis of the Kidney.

A kidney converted into a thin-walled multilocular sac and shewing its ureter to be blocked by a large soft black calculus.   The walls of the pelvis and of the upper part of the ureter are infiltrated with a deposit of malignant growth. On the reverse of the specimen the growth is seen to project into several of the dilated calices in the form of cauliflower-like excrescences which histologically have the structure of spheroidal-celled carcinoma.

From a patient æt. 71.

Presented by Mr. G. Parsons.

1704   Saccular Dilatation of the Ureter.

A left kidney from the hilum of which proceed three short ducts each opening into a spherical thin-walled cavity, which contained in the recent state a pint of urine. At the lower end of the cyst is seen the continuation of the ureter. The kidney is laid open and shews partial absorption of its pyramids. A similar condition of the upper part of the ureter is seen in the right kidney.

From a woman who was brought to the hospital dying from a dissecting aneurysm which opened into the sac of the pericardium. A Zwank's pessary was found in the vagina. *See Insp.* 1876, No. 153½.

1705   Dilated Ureters.

The ureters with the bladder laid open to shew the ducts dilated so as to measure an inch in circumference. Their walls are thin and the orifices in the bladder appear to be of normal size. There is slight hypertrophy of the bladder.

Alexander J., æt. 7, was admitted under Dr. Pye-Smith with membranous laryngitis for the relief of which tracheotomy was performed. On the following day he died, and at the autopsy the pelves of the kidneys were found to be extremely dilated; the prostate and urethra were healthy. *See Insp.* 1886, No. 208.

1706   Dilated Ureters.

The posterior wall of a bladder with the terminal portions of the ureters seen from behind. Both ureters are considerably dilated and their walls thickened. The right, which is the larger and admits the index finger, has a tortuous course and presents several incomplete septa in its lumen. The orifices of the ducts in the bladder appear to be normal.

1707   Dilated Ureter.

A portion of the base of a bladder with its ureters laid open to shew the right duct to be of normal size, whilst the left measures an inch and five-eighths in circumference. Nothing abnormal was discovered in the condition of the uretral orifices in the bladder.

Samuel S., æt. 48, was admitted under Dr. Hale White in a condition of alcoholic coma and died four days later from gangrenous pneumonia. At the autopsy two strictures were found in the urethra and the pelvis of the left kidney was dilated. The right kidney was unaffected. *See Insp.* 1887, No. 69.

1708   Dilated Ureter.   Hydronephrosis.

A kidney with its ureter mounted to shew the duct uniformly dilated and measuring an inch in diameter. The pelvis of the kidney is also enlarged and the organ itself is hydronephrotic.

Faith D., æt. 46, was admitted under Mr. Key in 1833 with symptoms of peritonitis from which three weeks later she died. At the autopsy a perforation was found in the posterior aspect of the bladder, the walls of which were infiltrated with cancer. *See Insp.* vol. 17, p. 97.

1709   Dilated Ureters.

The kidneys with the uterus and bladder mounted to shew the ducts dilated so as to measure an inch and a half in circumference at the widest part. The dilatation of the ureters is unequally distributed and their orifices in the bladder are wide and circular. The pelves of the kidneys are dilated and the organs themselves small and lobulated.

Edward C., æt. 15, was admitted under Dr. Bright in 1835 for hæmoptysis, œdema of the legs, and albuminuria. He died four weeks after admission, and at the autopsy the lungs were "generally affected with a recent chronic pneumonia in the form of a pale red hepatization, not very solid but tough and mostly devoid of air." There was general hypertrophy of the heart. *See Insp.* vol. 20, p. 33.

1710 Congenital Stenosis of the Ureter. Hydronephrosis.

A left kidney with the upper half of its ureter, at the commencement of which there is an almost impervious fibrous stricture about an eighth of an inch in length. The pelvis and infundibula are considerably dilated. The renal structure is normal.

Bertram K., æt. 8 months, was admitted under Dr. Perry for broncho-pneumonia, from which ten days later he died. At the autopsy the right kidney was found to be normal. *See Insp.* 1890, No. 68.

1711 Stenosis of the Ureter.

The upper part of a left ureter with a portion of the dilated pelvis of the kidney. The ureter at its commencement is greatly narrowed so as barely to admit a small probe, and the passage is overhung by a valvular fold of mucous membrane. Behind is mounted the right kidney, extremely atrophied and measuring an inch in length, together with its ureter, artery, and suprarenal capsule.

John R., æt. 33, was admitted under Dr. Barlow with a tumour in the left side of the abdomen, associated with the passage of large quantities of urine having a low specific gravity and containing a trace of albumen. Uræmia supervened, and he died in convulsions eleven days after admission. At the autopsy the left kidney was converted into a hydronephrotic sac which contained about 2 pints of limpid urine. *See Guy's Hosp. Reps.* 1845, p. 313.

1712 Ureter obstructed by Growth.

A portion of a left ureter which is somewhat tortuous, and is imbedded for the greater part of its length in a mass of growth. The tube has been laid open, and shews several constrictions, above the uppermost of which the lumen is dilated. Histologically the growth is a spheroidal-celled carcinoma with scanty stroma.

James G., æt. 74, was admitted under Mr. Durham for an epithelioma involving the base of the tongue and the submaxillary glands, symptoms of which had been noticed for five months. Erysipelas supervened, and the patient died two months after admission. At the autopsy secondary deposits were found in the lungs and in the bronchial and retro-peritoneal glands. The left kidney was in a condition of hydronephrosis. *See Insp.* 1804, No. 134.

1713   Pyonephrosis.

A left kidney converted into a globular sac about six inches in diameter. It has been laid open to shew its inner surface to be covered with granular lymph and to present several depressions corresponding to dilatations of the calices. The wall is about an eighth of an inch in thickness and histologically presents no trace of renal structure. The ureter is pervious and the bladder which is mounted below is much contracted, its mucous membrane being thickly coated with inflammatory deposit.

Mrs. S., æt. 34, was admitted under Dr. Addison in 1827 for a tumour on the left side of the abdomen which had been slowly growing for three years. Four months before her death her health failed and she was noticed to be passing scanty purulent urine. At the autopsy the cyst was found to contain a "dirty discoloured watery pus," and the right kidney, which appeared to be normal, was hypertrophied. *See Insp.* vol. 4, p. 117.

1714   Pyonephrosis.   Urethral Stricture.

A left kidney embedded in a thick layer of fibro-fatty tissue. The calices are dilated and in the recent state contained pus. The ureter is hypertrophied and enlarged.

Richard T., æt. 26, was admitted in 1826 for stricture of the urethra, and died suddenly from œdema of the glottis. At the autopsy the bladder was found to be inflamed, and the soft palate was perforated as the result of syphilitic ulceration. *See Insp.* vol. 11, p. 151; and *Prep.* 482.

1715   Calculous Pyonephrosis.

A right kidney enlarged so as to measure six inches in length and containing in the recent state about half a pint of pus. It has been laid open to shew the lining membrane of its dilated calices to be in parts smooth and fibrous, whilst elsewhere it is coated with a thin layer of granular lymph. The thin shell of cortical tissue is surrounded by fibro-fatty tissue to which the capsule is unduly adherent. The ureter is small and much thickened.

Anne L., æt. 29, was admitted under Dr. Bright in 1820 for a tumour about the size of one's fist situated in the region of the right kidney. The urine was thick and coagulated on the application of heat. She died from cynanche tonsillaris one month after her admission, and at the autopsy a small calculus was found in the pelvis of the right kidney. The left kidney was normal. See Insp. vol. 14, p. 92.

1716   Calculous Pyonephrosis. Perinephric Abscess.

A right kidney with its perinephric tissue and the upper part of the ureter. The organ has been laid open to shew the pelvis and calices somewhat dilated, and its cortical tissue occupied by several abscesses varying from a line to a third of an inch in diameter. The capsule is partially detached and upon the lower end of the kidney is seen a considerable mass of blood-clot. On the reverse of the specimen the upper part of the ureter is observed to be occupied by a black fusiform calculus.

Henry W., æt. 13, was admitted under Dr. Bright in 1820 for peritonitis, from which five days later he died. At the autopsy there was evidence of recent peritonitis and the right kidney was found lying in a collection of pus which separated the organ from its capsule. The left ureter contained a calculus and both pelves were distended with pus. See Insp. vol. 8, p. 163.

1717   Calculous Pyonephrosis. Nephro-colic Fistula.

A left kidney in a condition of hydronephrosis, the pelvis and some of the calices being filled with calculi.

At the lower end of the organ there is a sinus indicated by a yellow rod whereby the pyonephrosis communicated with the colon. Firmly attached around the opening is seen a small ring of intestinal mucous membrane.

Benjamin W., æt. 58, was a patient under the care of Dr. Barlow in 1835 for pain in the lumbar region associated with pyuria and hæmaturia. "His stools frequently contained pus, but his urine became more copious and clearer though his wife persisted in asserting that it was nearly all passed per anum." At the autopsy a calculus was found in the pelvis of the right kidney, which was hydronephrotic. There were tubercles in the lungs. *See Insp.* vol. 19, p. 131; and *Guy's Hosp. Reps.* 1842, p. 9.

1718   Calculous Pyelitis. Nephrectomy.

A kidney, contracted and cirrhotic, laid open to shew its calices dilated and converted into ragged-walled cavities, in one of which is lodged a small calculus. The surface of the organ is lobulated and nodular ; its capsule was thickened and firmly adherent.

The kidney was removed from a boy æt. 11 in 1888.

1719   Calculus in the Pelvis of the Kidney.

A left kidney in the pelvis of which is lodged a mulberry calculus. The organ is greatly contracted, measuring only two and a half inches in its longest diameter. It has been incised to shew its pelvis and calices dilated, and its substance occupied by numerous cysts.

George M., æt. 65, was admitted under Mr. Bransby Cooper in 1828 for a fracture of the right thigh. Several days later he died, and at the autopsy the right kidney "was far advanced in the white mottling degeneration." *See Insp.* vol. 6, p. 60.

1720   Calculus in the Kidney.

The half of a left kidney, the pelvis of which is occupied and the ureter occluded by a black moulded calculus. Several smaller calculi are seen in the calices. The structure of the kidney appears normal.

John B., æt. 22, was admitted under Dr. Curry in 1804 for pain and distension of the abdomen, vomiting, and suppression of urine. At the autopsy an intussusception was found at the lower end of the ileum. *See Old Museum Book*, no. 9.

1721   Renal Calculus.   Contracted Kidney.

A small kidney laid open to shew its pelvis and calices occupied by several rough white calculi. The surface of the organ is fissured and coarsely granular.

1722   Calculous Pyelitis.   Perinephritis.

A kidney embedded in and firmly united to its perinephric fat. It has been divided to shew several calculi lodged in its calices. The interior of the pelvis is roughened, and on the reverse of the specimen is seen the opening left after nephrolithotomy.

John B., æt. 36, was admitted under Mr. Durham for a sinus in the region of the right kidney which had existed for about five years. Twelve years previously a vesical calculus was removed by lithotomy. A month after admission the kidney was explored and seven stones removed. A week later he died from peritonitis consecutive to suppuration around the kidney extending downwards into the pelvis. *See Insp.* 1884, no. 148.

1723   Calculus in the Kidney.

A left kidney the pelvis of which is somewhat dilated and contains a friable phosphatic calculus. Attached to the kidney by the ureter is a portion of the bladder with the prostate. The prostate is enlarged and in the recent state the bladder, which is hypertrophied, contained a large calculus.   *Presented by Mr. Camplin.*

1724   Renal Calculi.   Atrophied Kidney.

A kidney much atrophied and containing several calculi. The pelvis is dilated and a stone is lodged at the orifice of the ureter. The renal artery is very small.

1725   Calculi in the Kidney.   Nephrectomy.

A right kidney removed by operation and divided

longitudinally, in the interior of which are seen large spaces formed by dilatation of the calices. They are filled by branching calculi, one of which occupies the pelvis of the organ and occludes the orifice of the ureter. The calculi are composed of urates and oxalates and some of them are encrusted with phosphates.

Fanny F., æt. 37, was admitted under Mr. Lucas in 1886 with symptoms of stone in the right kidney for which nephrectomy was performed. Four months later, a calculus which was producing suppression of urine was removed from the pelvis of the left kidney. The patient was in good health four years after the second operation. *See Surg. Reps.* vol. 126, Case 38 A.

1726   **Calculus impacted in the Ureter.**

The lower end of the right ureter with a portion of the bladder, vesiculæ seminales, and prostate. About one inch from the orifice of the ureter is lodged an oval calculus measuring a third of an inch transversely. Above the calculus the ureter is somewhat dilated.

1727   **Phosphatic Deposit in the Renal Pelvis.**

The half of a kidney, the pelvis and ureter of which have been laid open and are seen to be partially covered by a deposit of phosphatic material.

Presented by Mr. Hilton.

1728   **Renal Calculus.**

A minute calculus measuring an eighth of an inch in diameter and divided to shew a nucleus consisting of blood-clot surrounded by phosphatic deposit. It was passed with several similar concretions by a patient suffering from malignant disease of the kidney.

Presented by Dr. Owen Rees, 1883.

1729   **Renal Calculi.**

A series of renal calculi, the smallest of which are facetted and roughly pyramidal, whilst others are spiculated. The largest is nodular and ovoid on section.

They are white in colour, the largest having a brown exterior and on section presenting a laminated appearance.

Robert G., æt. 21, was admitted under Dr. Bright in 1840 for abdominal pain of seven months' duration. He died three weeks after admission, and at the autopsy the right kidney was hydronephrotic, and contained the calculi which form the preparation. *See Insp.* vol. 30, p. 315.

1730   Branched Renal Calculus.

A very large calculus, presenting three main branches corresponding to the infundibula of the kidney, and several smaller excrescences moulded to the dilated calices. The main branches are, for the most part, smooth, and the excrescences rough and nodular.

Presented by Dr. Bright.

1731   Branched Renal Calculus.

A large branched calculus moulded to the cavity of a hydronephrotic kidney. The surface of those parts which lie in the dilated calices is white and covered with short spicules, while elsewhere the calculus is smooth and has a brownish tint. It has been analysed by Dr. Stevenson, who found its outer layer to be composed of pure triple phosphate, and its interior to consist of a mixture of triple and calcium phosphate, with a very small proportion of uric acid. The calculus weighs 1650 grains. *See Trans. Path. Soc.* 1885, p. 276.   *Presented by Dr. Hale White.*

1732   Renal Calculus.   Nephro-lithotomy.

A calculus weighing 18 grains, resembling in shape a suprarenal body, and measuring three quarters of an inch in its longest diameter. Its surface is rough and of a brownish colour, whilst its interior is white.

The stone was removed from the renal pelvis of a patient 50 years of age, by Mr. Symonds in 1883.

1733   Renal Calculi.

A considerable number of renal calculi, varying greatly in size and shape, the smallest being about a line in diameter. The larger ones are branched, and moulded to the calices of the kidney. They have a rough surface and are white in colour. Chemical examination shews them to consist mainly of oxalate of lime with traces of uric acid. *Presented by Mr. R. Stocker.*

1734   Renal Calculus.

A small dark brown calculus, the surface of which is covered with minute spicules. Chemically it consists of oxalate and phosphate of lime, with traces of magnesia.

John D., æt. 41, was admitted under Dr. Pye-Smith for chronic interstitial nephritis, from which he died. At the autopsy the calculus was found lying loose in the pelvis of the kidney. There was a deposit of urate of soda in the great-toe joints. *See Insp.* 1886, No. 37.

1735   Renal Calculus.

The half of a mulberry calculus, the section of which is oval and measures one inch in its longest diameter. It has a laminated structure. It was analysed by Dr. Golding Bird, and found to have a specific gravity of 1·85, and to consist of oxalate of lime.

Mary Anne F., æt. 42, was admitted under Dr. Ashwell in 1839 for malignant disease of the liver, from which, one month later, she died. At the autopsy the left kidney was found to be four times its natural size and in a condition of hydronephrosis. The stone which forms the preparation was lodged at the commencement of the ureter. *See Insp.* vol. 28, p. 114.

1736   Phosphatic Concretions from the Kidney.

White fragments and sand-like matter found in the pelvis of a kidney. The deposit was analysed by Dr. B. Babington, and consists of phosphate of lime.

From John P., æt. 53, who at intervals during the four years preceding his death passed blood and "earthy matter" with his urine. At the autopsy the right kidney was filled with soft cretaceous material; there was chronic cystitis. The mucous membrane of the pelvis of the left kidney was inflamed and partially encrusted by material similar to that forming the preparation. *See Prep.* 2030; and *Note Book*, i. p. 154.

Presented by Mr. E. C. May, 1830.

1737   Renal Calculus.

A nodulated calculus having a rough brown surface and white interior, which has been found on analysis to consist of phosphate of magnesium and ammonium. It was removed from the pelvis of the kidney.

Presented by Dr. Hale White.

1738   Thrombosis of the Renal Veins. Lardaceous Disease.

The half of a kidney, the cut surface of which shews the renal veins to be filled with thrombus. The surface of the organ is smooth, and histologically the glomeruli and arterioles shew lardaceous change.

Thomas McG., æt. 33, was admitted under Mr. Bryant for disease of the femur of twenty years' standing. The bone was resected and the patient died two weeks later. At the autopsy the kidneys were found to weigh 27 ounces, and the lumbar glands were lardaceous. The liver and spleen appeared normal. *See Insp.* 1809, No. 312.

1739   Thrombosis of Renal Veins.

A kidney divided longitudinally to shew the branches of the renal vein distended with adherent thrombus. The kidney itself is enlarged so as to measure six inches in length, and histological examination shews advanced tubal and interstitial nephritis with lardaceous change in the glomeruli and arterioles.

1740   Thrombosis of Renal Vessels.

A section of a left kidney mounted to shew the renal vessels occupied by thrombus. Histologically the secreting substance appears normal, and the vessels, both arteries and veins, are filled with partially organised blood-clot.

Emily N., æt. 33, was admitted under Mr. Poland for injuries sustained by being crushed between a railway carriage and a platform. Six days later she died, and at the autopsy the right ureter was found to be ruptured, the processes of the lumbar vertebræ were fractured, and the inferior vena cava contained thrombus. *See Insp.* 1868, No. 25 A; and *Prep.* 1590.

1741   Atheromatous Renal Arteries.

A portion of a kidney partially dissected to shew the main branches of the renal artery thickened and atheromatous. Histologically the tissue of the organ is normal.

Charles S., æt. 65, was admitted under Mr. Symonds for gangrene of the foot, for which amputation was performed through the lower third of the thigh. Fourteen days later he died, and at the autopsy there was tuberculous disease of the lungs, larynx, and colon. The aorta was only slightly atheromatous, the smaller arteries being considerably diseased. *See Insp.* 1895, No. 179.

Section XXXIII.—INJURIES AND DISEASES OF THE URINARY BLADDER.

Malformations: 1742–1748.
Cyst of Urachus: 1749.
Rupture: 1750–1754.
Perforation: 1755–1757.
Hæmorrhage into wall: 1758, 1759.
Inflammation: 1760–1768.
Hypertrophy: 1769–1774.
Dilatation: 1775.
Fatty Infiltration: 1775.
Sacculation: 1776–1781, 1803.
Tubercle: 1782–1785.
Papilloma: 1786–1792.
Villous Tumour: 1793.
Epithelioma: 1794–1803.
Carcinoma: 1804–1817.
Sarcoma: 1818–1827.
Hydatid: 1828–1833.
Dermoid Cyst: 1834, 1835.
Perivesical Abscess: 1836, 1837.
Vesical Fistulæ: 1838, 1839.
Prolapsed Ureter: 1840.
Calculi in Bladder: 1841–1846.
Results of Lithotomy: 1847–1854.

1742   Ectopia Vesicæ.

The genito-urinary organs of an adult male with portions of the ossa innominata and of the abdominal wall. Immediately above the penis, which is short, devoid of prepuce, and curved downwards, is a shallow depression representing the posterior wall of the bladder covered by mucous membrane closely resembling the surrounding skin. The openings of the ureters into this depression are
.

indicated by yellow rods, and a little anterior to these is an opening, through which a rod has been passed to the right vesicula seminalis. The pubic bones are separated from each other for a distance of about two inches. The testes and vasa deferentia are well developed.

Presented by Alexander Bossy, Esq.

1743   Ectopia Vesicæ.

The genito-urinary organs of a male fœtus with a portion of the anterior abdominal wall, mounted to shew immediately above the pubes a small opening about a third of an inch in diameter, through which is seen the interior of the bladder. Above this orifice is situated the umbilical cord, whilst below it is a rudimentary penis. The kidneys and ureters appear to be normal.

1744   Extroversion of the Bladder.

A dissection of the parts concerned in ectopia vesicæ. Immediately above the pubes the anterior wall of the bladder is seen to be deficient and its extroverted posterior wall is covered by a thickened warty mucous membrane. Below the orifices of the ureters, marked by red rods, is seen the glans of a stunted penis with a partial prepuce limited to its under surface.

David D., æt. 5 months, was admitted under Mr. Bryant for epispadias and extroversion of the bladder. A plastic operation was performed and sixteen days later the child died from suppurative peritonitis. There was a congenital stricture of the rectum. *See Insp.* 1886, No. 98 ; and *Prep.* 983.

1745   Extroversion of the Bladder.

A portion of the anterior abdominal wall presenting at its lower part, immediately above the pubes, an opening through which protrudes the extroverted bladder. The viscus forms a flat oval swelling measuring an inch and a quarter transversely in its longest diameter. At its lower margin are the openings of the ureters, and immediately below it is a small rudimentary penis. On either side are the incisions made with a view of forming an anterior wall to the bladder. The ureters and the pelves of the kidneys are somewhat dilated.

From Robert G., æt. 14 months, who died shortly after the performance of a plastic operation.

1746   Extroversion   of   the   Bladder.   Plastic operation.

An extroverted bladder with the adjacent parts mounted to illustrate the character of a plastic operation performed for the relief of the condition. Flaps were dissected upon either side of the bladder and sewn together, the testes were excised and the penis passed through a hole in the scrotum, the scrotum itself being brought up to complete the anterior wall of the viscus.

James II., æt. 31, was admitted under Mr. Symonds with extroversion of the bladder, for the relief of which a plastic operation was performed. Four days later the patient died from cellulitis. At the autopsy the pubic bones were found to be separated from each other for a distance of an inch and a quarter. *See Insp.* 1892, No. 326.

1747   Extroversion of the Bladder.

The female genital organs together with the lower part of the abdominal wall, which is deficient in front and shews through the opening the extroverted mucous membrane of the posterior wall of the bladder. The mucous membrane is thickened and presents several rounded prominences, between the lowest of which open the dilated ureters. The clitoris is absent, the labia are large, and the hymen with the vagina and uterus are normal.

From a child, æt. 23 months, who died from an acute fever of about ten days' duration. At the autopsy the symphysis pubis was found to be deficient, the bones being an inch apart. *See Insp.* vol. 6, p. 93.

1748   **Extroversion of the Bladder.**

A bladder with the female genital organs and a portion of the abdominal wall. The anterior wall of the bladder and the corresponding portion of the parietes are wanting. The mucous surface of the posterior vesical wall forms an oval prominence, at the lower part of which are seen the orifices of the ureters. On the reverse of the specimen these ducts are seen to be dilated. The vagina is large and its hymen is imperforate.

From Brookes' Collection.

1749   **Cyst of the Urachus.**

The apex of a bladder with a portion of the urachus. Immediately above the bladder the urachus is distended into a small cyst measuring about a third of an inch in diameter. The wall of the cyst is thin and its lining is smooth.

1750   **Rupture of the Bladder.**

A bladder laid open to shew on its posterior wall, somewhat to the right side, a vertical laceration which in the recent state measured four inches in length and communicated freely with the peritoneal cavity.

An adult male was admitted under Mr. Poland with abdominal symptoms. Thirty-six hours previously, while intoxicated, he jumped a stile and fell violently upon a heap of stones. On the day following admission six pints of bloody fluid were withdrawn by catheter. Some hours later the patient died and at the autopsy the abdominal cavity contained about two quarts of bloody fluid, and there was early general peritonitis. *See Insp.* 1862, No. 162.

1751   **Rupture of the Bladder.**

A somewhat enlarged bladder shewing on its posterior wall a vertical laceration four inches in length extending as far as the summit of the organ.

John P., æt. 38, was admitted under Mr. Durham with severe abdominal pain and vomiting. Fourteen hours previously a heavy man had fallen upon him in wrestling, the patient having at the time " four hours' urine in his bladder." Shortly after admission a catheter was passed and about six pints of bloody urine were withdrawn. Three days later he died and at the autopsy a rent was found in the peritoneal coat of the ileum and there was much lymph in the recto-vesical pouch with evidence of recent general peritonitis. *See Insp.* 1873, No. 384.

1752   Ruptured Bladder.   Sutured.

The female pelvic organs with a portion of the anterior abdominal wall seen from behind, and shewing at the summit of the bladder a large ragged opening freely communicating with the peritoneal cavity. There is recent lymph upon the surrounding peritoneum.

Frances H., æt. 7, was admitted under Mr. Symonds in 1887, having been thrown by her lunatic mother from Blackfriars Bridge into a passing barge 40 feet below. Laparotomy was performed a few hours later and the rent in the bladder was sewn up. The patient died on the eighth day, and at the autopsy suppuration was found in the cellular tissue between the bladder and the anterior abdominal wall. *See Trans. Clin. Soc.* 1888, p. 228.

1753   Laceration of the Bladder.

A portion of a bladder laid open to shew on its postero-lateral wall, half an inch from the orifice of the right ureter, a perforation about an eighth of an inch in diameter.

Evan T., æt. 55, was admitted under Mr. Cock for injuries received from a cask of tallow falling upon him while working in a ship's hold. A catheter was passed and some bloody urine was withdrawn. Five days after admission the patient died from hypostatic pneumonia, and at the autopsy the pelvis was found to be fractured and the cellular tissue between the bladder and rectum was gangrenous. *See Insp.* 1855, No. 209.

1754   Laceration of the Bladder.

A bladder with the anterior portion of the bony pelvis mounted to shew a large ragged opening obliquely crossing the summit of the organ. The right pubic bone has been separated from the cartilage of the symphysis.

Thomas K., æt. 44, was admitted under Mr. Durham, having been run over by a timber cart, which passed across his abdomen. There was abnormal mobility of the pelvic bones and the urine contained blood. Four days after admission he died from general peritonitis. The autopsy revealed numerous fractures of the pelvis. *See Insp.* 1870, No. 486.

1755   **Bladder perforated by a Catheter.**

A bladder everted to shew upon its posterior wall, midway between the base of the trigone and the summit of the organ, a small circular opening which communicated with the cellular tissue between the bladder and rectum.

David J., æt. 32, was admitted under Mr. Cock, having been struck by a piece of falling timber. His injuries proved fatal three days later. A catheter was passed on several occasions and it was thought that the perforation in the bladder was produced by this instrument. At the autopsy the liver was found to be lacerated and the pubic bones were widely separated. *See Insp.* 1855, No. 210.

1756   **Bladder perforated by a Catheter.**

A portion of a bladder shewing on its posterior wall, about an inch from the orifices of the ureters, two small rounded perforations, one of which communicates with the peritoneal cavity, whilst the other leads into a suppurating sac situated beneath the serous coat of the organ.

From a patient who was in the habit of passing a catheter on himself.

Presented by Mr. Hilton.

1757   **Bladder perforated by a Catheter.**

A sagittal section through the bladder and rectum shewing on the posterior wall of the former a small rounded perforation, through which a blue rod has been passed into the recto-vesical pouch. There is recent lymph upon the peritoneum covering the bladder.

Thomas G., æt. 18, was admitted under Mr. Davies-Colley for sinuses in the groin and in the perinæum consequent upon lithotomy performed seven years before admission. The bladder was opened by Cock's operation and a few days afterwards symptoms of peritonitis supervened. It was thought that the perforation in the bladder was caused by the catheter which had been tied in at the operation. *See Insp.* 1881, No. 270; and *Prep.* 1852.

1758   Hæmorrhage into the Wall of the Bladder.

A bladder laid open to shew its mucous membrane extremely rugose and raised into nodular prominences. The cut surface of its wall demonstrates that the condition is due to extensive infiltration of blood into the submucous and muscular coats.

James M., æt. 21, was admitted under Dr. Gull with symptoms of meningitis of a fortnight's duration. His urine was withdrawn by catheter and was found at first to be free from albumen, and on the day of his death, six days after admission, to contain much blood. He also suffered from hæmatemesis. At the autopsy old tuberculous disease was found in the lungs and peritoneum and there was acute general tuberculosis. *See Insp.* 1857, No. 237; and *Drawing*, 367 (20).

1759   Blood-Tumour in the Bladder.

A female bladder, at the neck of which is seen, on its posterior surface, a rounded elevation a third of an inch in diameter covered by normal mucous membrane, and presenting a slight constriction at its point of attachment. Histological examination shews that the tumour consists of blood-clot lying in the submucous and muscular coats of the organ.

Mary M., æt. 63, was admitted under Mr. Lucas with a strangulated femoral hernia, for the relief of which herniotomy was performed. She died about five weeks after admission, and at the autopsy the mucous membrane of the bladder was of a bright red colour and the kidneys were in a condition of suppurative pyelo-nephritis. *See Insp.* 1893, No. 228.

1760   **Acute Cystitis.**

A contracted bladder with the proximal portion of
the urethra laid open to shew the mucous membrane
coated with a thick adventitious deposit, which, in the
urethra, is continuous and in the bladder has a patchy
distribution and is thickest on the left side. The
ureters are lined with a similar material. Histological
examination of the wall of the bladder shews superficial
coagulation necrosis with thrombosis of the vessels.

James B., æt. 32, was admitted under Mr. Poland with
symptoms of acute cystitis attributed to catching cold. He
suffered great pain in the lower part of the abdomen and died
twenty-five days after admission. At the autopsy multiple
abscesses were found in the kidneys. *See Insp.* 1863, No. 261.

1761   **Acute Cystitis.**

A bladder laid open to shew its mucous surface
markedly rugose and thickly coated with false membrane. In the recent state it was deeply congested
and in parts blackened. On the reverse of the specimen
the serous coat of the organ is seen to be covered by
a thin layer of lymph. Histological examination shews
that the inflammatory process has led to the destruction
of the mucous membrane and has affected all the coats
of the organ.

Sophia B., æt. 61, was admitted under Mr. Howse for
abdominal pain and distension. A catheter was passed and the
urine was found to contain pus. Four days later the urine
contained blood ; and the patient died twelve days after
admission. At the autopsy coils of small intestine were found
adherent to the bladder, and the kidneys were in a condition of
pyelo-nephritis. *See Insp.* 1892, No. 70.

1762   **Acute Cystitis.**

The urinary organs of a female infant mounted to
shew the walls of the bladder to be thickened and its
interior lined by a false membrane, which can be

traced into the lower ends of the ureters. These ducts are considerably dilated and in the recent state the pelves of the kidneys were reddened and inflamed, and the apices of the pyramids were necrotic.

Mary C., æt. 9 months, was admitted under Dr. Goodhart in a moribund condition. The child had been brought up at a Baby-farm and was blind in both eyes from corneal opacity. At the autopsy small patches of broncho-pneumonic consolidation were found in the lower lobe of the left lung. See Insp. 1893, No. 56.

1763   Membranous Cystitis.

The posterior half of a bladder mounted to shew upon its interior, a little above the orifice of the ureters, a patch of membranous exudation firmly adherent to the mucous surface. There is considerable hypertrophy of the muscular coats of the organ. The ureters are much dilated.

Edward P., æt. 25, was admitted under Mr. Bransby Cooper in 1837 with symptoms of stone in the bladder, from which he had suffered for about eighteen years. Lithotomy was performed and three days later the patient died from suppurative peritonitis. At the autopsy the kidneys were found to be hydronephrotic, and the cellular tissue about the neck of the bladder was soft and gangrenous. See Insp. vol. 2, p. 68.

1764   Phlegmonous Cystitis.

An enlarged bladder laid open to shew its interior covered by false membrane encrusted with phosphatic deposit. The walls of the organ are much thickened by purulent infiltration.

John O., æt. 36, was admitted under Mr. Morgan in 1836 for a perinæal abscess following upon stricture of the urethra. A week later he died and at the autopsy the urethra was found to be lined with false membrane and there was suppurative nephritis. See Insp. vol. 23, p. 28.

1765 **Sloughing of Mucous Membrane of the Bladder.**

A large female bladder with thickened walls everted to shew its mucous membrane almost entirely detached, exposing the subjacent muscular coat. The mucous coat shews numerous perforations and is thickened by phosphatic deposit.

Eliza S., æt. 32, was admitted under Mr. Durham for retention of urine, having suffered from vesical symptoms for five months. On admission a large quantity of fœtid blood-stained urine was drawn off, and fifteen days later a piece of sloughing mucous membrane was removed through the dilated urethra. Three days after the operation the patient died, and at the autopsy the ureters were found to be full of pus, and there were multiple abscesses in one of the kidneys. *See Insp.* 1878, No. 239.

1766 **Cystitis in Paraplegia.**

A bladder laid open from the front to shew its mucous membrane thickened and rugose. In each lateral wall of the organ there is a large ragged opening, which in the recent state communicated with abscess cavities in the connective tissue. At the apex the cellular tissue between the bladder and the anterior abdominal wall is seen to be shreddy and gangrenous. On the reverse of the specimen a coil of small intestine is attached by recent lymph to the serous coat of the viscus. Histologically all the coats of the organ are seen to be infiltrated with small round cells.

James H., æt. 21, was admitted under Mr. Aston Key in 1827 for fracture of the lower dorsal spine, producing complete paraplegia. The patient survived the injury four weeks, during which period catheterism was daily employed. At the autopsy there was general purulent peritonitis. *See Insp.* vol. 4, p. 55.

1767 **Cystitis in Paraplegia.**

A bladder opened by removal of its anterior wall to shew the mucous membrane over the lower half of the organ in some parts covered by false membrane and in others removed by ulceration. The wall of the viscus is thickened, and histological examination shews that the cellular tissue between the muscular bundles is œdematous and infiltrated with small round cells.

Frederick H., æt. 20, was admitted under Mr. Bransby Cooper in 1828 for paraplegia, attributed to an injury received six months previously. While in the hospital the bladder was periodically emptied by catheters and ten weeks after admission the patient died. At the autopsy the spine was found to be affected by malignant disease and there was ascending nephritis. *See Insp.* vol. 4, p. 64 ; and *Prep.* 1037 [2nd Edit.].

1768   Cystitis in Paraplegia.

The posterior half of a bladder, the mucous membrane of which presents a granular surface from the presence of false membrane and is in parts eroded by ulceration. There is recent lymph upon the peritoneal coat, and in the recent state the walls of the organ were in a condition of diffuse purulent infiltration.

James P., æt. 42, was admitted under Mr. Davies-Colley for fracture of the lower dorsal spine producing complete paraplegia. The urine was withdrawn by catheter from the first, and four days after admission it was found to be alkaline and to contain blood. The patient died ten days after the accident, and at the autopsy a coil of intestine was found to be adherent to the summit of the bladder and there was acute pyelitis. *See Insp.* 1881, No. 101.

1769   Hypertrophied Bladder.   Stricture of Urethra.

A bladder with the proximal half of the urethra laid open to shew the muscular coat of the viscus hypertrophied, so as to measure three quarters of an inch at its thickest part. There is a stricture at the bulbous portion of the urethra.

John W., a middle-aged man, was admitted under Sir Astley Cooper in 1804 with symptoms of stone in the bladder. He died a fortnight later, and at the autopsy the bladder was found to be entirely filled by a calculus and the kidneys were in a condition of suppurative nephritis. *See Old Museum Book*, No. 67 ; and *Prep.* 2198 [2nd Edit.].

1770   **Hypertrophied Bladder. Cystitis.**

A bladder laid open from the front to shew its muscular coat considerably thickened and its mucous membrane rugose and blackened by inflammation.

John C., æt. 37, was admitted under Mr. Morgan in 1836 for retention of urine, and died two months later. At the autopsy the skin of the penis and perinæum was sloughing as a result of extravasation of urine. The prostate was enlarged and the ureters and pelves of the kidneys were dilated. *See Insp.* vol. 21, p. 101.

1771   **Hypertrophy and Fasciculation of the Bladder.**

A portion of a hypertrophied bladder everted to shew fasciculation of its interior. The interlacing bundles of muscular tissue are greatly hypertrophied and between them are numerous depressions.

Daniel W., æt. 49, was admitted under Mr. Cock for a urethral stricture of twenty-six years' duration, which had recently given rise to a perinæal abscess. The abscess was opened, but pyæmia supervened and the patient died six weeks after admission. At the autopsy the stricture was found to be situated in the membranous and spongy portions of the urethra and there were many false passages. There was septic broncho-pneumonia. *See Insp.* 1855, No. 26.

1772   **Hypertrophied Bladder.**

A bladder, the cavity of which is large and its walls somewhat thickened. Its interior is fasciculated.

James B., æt. 50, was admitted under Dr. Pavy for diabetes mellitus, and died a month later from lobar pneumonia. At the autopsy the kidneys were found to be rough on the surface and to weigh twelve ounces. The urethra was normal. *See Insp.* 1874, No. 457.

1773   **Hypertrophied Bladder.**

A bladder divided by a sagittal section and mounted to shew considerable hypertrophy of its muscular coat. The hypertrophy is most marked at the apex and along the anterior wall.

Samuel R., æt. 44, was admitted under Dr. Pavy for albuminuria and œdema of the legs of a few weeks' duration. He
had long suffered from a stricture of the urethra. A few days
after admission he died from urœmic convulsions, and at the
autopsy the stricture was found to be situated in the membranous
portion of the urethra, the kidneys were coarsely granular, and
the heart weighed twenty ounces. *See Insp.* 1838, No. 438.

1774   Hypertrophied Bladder.

A child's bladder, the cavity of which is of natural size,
whilst its wall is hypertrophied so as to measure three-
eighths of an inch in thickness. The mucous membrane
appears to be healthy.

Charles W., æt. 6, was admitted under Mr. Davies-Colley
with symptoms of stone in the bladder. A uric acid calculus
weighing thirteen grains was removed by lateral lithotomy. Twelve
days later the child died, and at the autopsy the right pleural
cavity was found to contain fourteen ounces of fluid, and the lungs
were affected with bronchiectasis and lobular pneumonia. *See Insp.*
1892, No. 135.

1775   Dilated and Fatty Bladder.

A bladder measuring eight inches in its longest
diameter, and presenting considerable thickening of
its walls. Its interior presents numerous shaggy
processes. The muscular coat is softened and appears
to have undergone extreme fatty degeneration.

From a man, æt. 50, who died from erysipelas, following the
removal of a small tumour from the eyelid. The patient had
suffered from paraplegia for five years, and had had difficulty in
micturition, so that he would sometimes retain his urine for a
whole day. He was extremely obese, and at the autopsy four
inches of fat were found on the abdominal wall. *See Note Book,*
p. 146; and *Preps.* 1668 (32) [2nd Edit.] and 1219.

Presented by Mr. Hilton, 1829.

1776   Sacculus of the Bladder.

A portion of a bladder showing on its posterior wall,
about three quarters of an inch from the orifice of the

right ureter, a thin-walled globular sacculus about as large as a pigeon's egg.

Presented by Sir Astley Cooper.

1777   Sacculus of the Bladder.

A hypertrophied bladder, to the right lateral wall of which is attached a sacculus, having a capacity nearly twice as great as that of the viscus itself. The two cavities communicate with each other by a smooth circular opening two-thirds of an inch in diameter, situated close beside the orifice of the right ureter. The cyst has a thin wall, and its lining is continuous with the mucous membrane of the bladder. Both ureters are dilated, and several false passages are seen at the neck of the bladder on the right side.

1778   Sacculus of the Bladder.

A bladder opened by the removal of its anterior wall and shewing on its left side a globular sacculus measuring two and a half inches in diameter, which communicates with the bladder by an oval aperture three-quarters of an inch in length. Histologically the wall of the sac is seen to be composed of mucous membrane and fatty tissue with a few bundles of muscular fibre. There is a second small sacculus on the posterior wall about half an inch from the orifice of the right ureter. The coats of the bladder are considerably hypertrophied.

James G., æt. 56, was admitted under Mr. Howse for retention of urine, having suffered from pyuria for four years. He died three weeks after admission, and at the autopsy the prostate was found to be enlarged and to contain abscesses. No stricture of the urethra was detected. *See Insp.* 1885, No. 280.

1779   Sacculated Bladder.

A bladder, the anterior wall of which has been removed to shew on its posterior surface three circular openings, measuring about a quarter of an inch in diameter. These openings communicate with sacculi, the largest of which is seen on the reverse of the specimen to measure an inch in its longest diameter, and to be partially covered by muscular tissue. At the apex of the organ is another sacculus as large as a pea, the thin wall of which is formed by a protrusion of the mucous membrane through the muscular coat.

Thomas H., æt. 51, was admitted under Mr. Davies-Colley for a perinæal fistula, having suffered from difficulty in micturition for twenty-four years. He died from sudden cardiac failure three days after admission. At the autopsy a tight stricture was found in the bulbous portion of the urethra; the kidneys were scarred and the heart weighed 22 ounces. *See Insp.* 1884, No. 294 ; and *Prep.* 530.

1780   Sacculated Bladder.

A bladder laid open to shew upon its posterior wall the orifices of numerous sacculi. On the reverse of the specimen these sacculi are seen to vary in size from a pea to a walnut, some having walls consisting only of mucous membrane, whilst in others the mucous membrane is supported by bands of muscular tissue. The muscular coat of the viscus is hypertrophied.

James A., æt. 35, was admitted under Mr. Durham with retention of urine, for the relief of which Cock's operation was performed. Six weeks later a swelling appeared at the lower part of the right chest from which pus was evacuated. He died thirty days after the operation, and at the autopsy a tight stricture was found in front of the membranous urethra, and the kidneys were in a condition of acute suppurative nephritis. There was a large single abscess in the right lobe of the liver. *See Insp.* 1885, No. 196.

1781   Sacculated Bladder.   Perforation.

A hypertrophied bladder, the interior of which is fasciculated and sacculated. At the apex of the organ is an aperture which in the recent state communicated with the peritoneal cavity and was thought to be due to traumatic rupture of a small sacculus. The prostate is greatly enlarged and its middle lobe, which forms a pyriform prominence at the neck of the bladder, has been perforated by a catheter. The membranous portion of the urethra presents a large opening communicating externally with an incision in the perinæum.

Joseph S., æt. 62, was admitted under Mr. Bryant for frequent micturition. Lateral lithotomy was performed and a small phosphatic calculus was removed. Cellulitis supervened and eleven days later the patient died. At the autopsy suppurative peritonitis was discovered and there were multiple abscesses in the kidney. *See Insp.* 1874, No. 469.

1782   Tuberculosis of the Bladder.

A bladder, beneath the mucous membrane of which project several rounded nodules varying in size from a line to a quarter of an inch in diameter. The right ureter at its orifice is thickened, and there are numerous areas of superficial ulceration, some circular and others irregular in shape, scattered throughout the interior of the bladder.

John J., æt. 22, was admitted under Dr. Gull with cerebral symptoms from which three days later he died. At the autopsy miliary tubercles were found in the meninges, lungs, liver, and spleen. The kidneys contained caseous tubercles, and the pelvis and ureter on the right side were ulcerated. *See Insp.* 1863, No. 52.

1783   Tuberculous Ulceration of the Bladder.

A bladder laid open to show between the orifices of the ureters a triangular ulcer, in the base of which are seen nodules of caseous material. Elsewhere similar nodules of about the size of millet seeds project beneath the mucous membrane, some of them presenting superficial ulceration. In the membranous portion of the urethra are several sinuses and the anterior part of the tube is deeply ulcerated. On the reverse of the specimen the ureters and vesiculæ seminales are seen to be infiltrated with caseous deposit.

o 2

Henry G., æt. 32, was admitted under Dr. Wilks with symptoms of pulmonary disease, having previously suffered from a perinæal abscess. He died about three weeks after admission, and at the autopsy tuberculous disease was found in the lungs, intestines, and kidneys. *See Insp.* 1859, No. 103.

1784   Tuberculous Ulceration of the Bladder.

A bladder shewing at the upper part of its posterior wall in the median line a distinct oval ulcer one-third of an inch in its longest diameter. Around the orifice of the left ureter is an irregular area of superficial ulceration. On the reverse of the specimen the ureter is seen to be thickened by a deposit which has the histological characters of caseous tubercle.

Reuben B., æt. 10, was admitted under Mr. Symonds for hip-disease. Eight months after admission amputation was performed and the patient died about twenty-four hours later. At the autopsy the spleen, liver, and kidneys were found to be lardaceous, and there was tuberculosis of the lungs and left kidney. *See Insp.* 1884, No. 96.

1785   Tuberculous Ulceration of the Bladder.

A bladder considerably dilated and laid open to shew around the orifice of the left ureter a deposit of caseous material, oval in shape, and measuring two and a half inches in its longest diameter. The margin of the deposit projects in parts considerably above the healthy mucous membrane, and its surface is irregular from ulceration. Two similar patches encrusted with phosphates exist at the summit of the organ. The left ureter is seen to be greatly thickened by a deposit having the histological characters of caseating tubercle. The prostate is also tuberculous.

James D., æt. 38, was admitted under Dr. Wilks with signs of phthisis associated with pigmentation of the skin, and died eight weeks later. At the autopsy tuberculous excavations were found in the lungs, with ulceration of the larynx and intestine. The adrenal bodies were caseous, and the left kidney was in a condition of tuberculous pyonephrosis. *See Insp.* 1884, No. 128.

1786   Papillomata of the Bladder.

A bladder laid open in front, to shew upon its interior surface several tufts consisting of long, slender, villous processes. A section of the wall of the bladder shews that the growths are attached to the mucous membrane, and do not invade the deeper coats. Histologically they are papillomata.   *From Brookes' Collection.*

1787   Papilloma of the Bladder.

A portion of the bladder shewing a button-like mass of warty growth, an inch in diameter, which is attached by a short narrow pedicle to the mucous membrane near the orifice of the left ureter. The interior of the bladder is much ulcerated and encrusted with phosphatic deposit. Histologically the tumour is composed of closely packed villous processes attached to a central core of well-formed fibrous tissue permeated by large vessels.

William J., æt. 72, was admitted under Mr. Bransby Cooper in 1836 for hæmaturia. He died of cystitis and suppurative nephritis. *See Insp.* vol. 21, p. 63; *Drawing*, 869; and *Guy's Hosp. Reps.* vol. i. p. 204.

1788   Papillomata of the Bladder.

A bladder opened in front to shew its interior covered with large tufts of villous growth, which are most abundant on the posterior wall and at the orifices of the ureters. They are composed of long delicate fimbriæ springing from the mucous membrane, and their tips are coated with phosphatic deposit. The wall of the bladder is much hypertrophied and the ureters are dilated. Microscopically the growths have the structure of papillomata.

From William N., æt. 50, who suffered from hæmaturia at intervals during the last six years of life.

Presented by Dr. Gull, 1861.

1789   Papilloma of the Bladder.

A papillomatous tumour having a stout pedicle, two inches in length, and covered with delicate villous processes.   It was removed from the bladder by operation.

Henry W., æt. 32, was admitted under Mr. Davies-Colley for hæmaturia and frequency of micturition. These symptoms had existed eight years. Perineal section was performed, and the growth removed. Four years after the operation the patient was in good health. *See Surg. Reps.* vol. 86, No. 26 ; and *Trans. Clin. Soc.* 1881, p. 104.

1790   Papilloma of the Bladder.

Tufts of simple papillomatous growth removed from the bladder by operation.

George P., æt. 55, was admitted under Mr. Bryant for hæmaturia and frequent micturition. The hæmaturia was first observed five years previously, had recurred at long intervals, and a month before admission became continuous. He died of peritonitis and pelvic cellulitis three days after perineal section had been performed. *See Insp.* 1884, No. 183.

1791   Papilloma of the Bladder.

A bladder opened in front shewing at its apex a small area of its mucous membrane covered by short papillomata.   In the recent state the bladder contained a straw coated with phosphates, one end of which projected into the prostatic urethra, and the other impinged upon the diseased surface.

From a man, aged 60, who died of leprosy. His chief symptoms during the last four months of life were referable to cystitis. It is supposed that the disease was induced by the irritation of the foreign body. *See Trans. Path. Soc.* vol. 38, p. 545 ; and Prep. 2013.

Presented by Dr. Beaven Rake, 1886.

1792   Papilloma of the Bladder.

A bladder laid open, shewing an oval tumour which was removed by operation and has been replaced.   The growth, which measures one inch and three quarters in its longest diameter, was situated just above the orifice of the right ureter, and had a constricted attachment to the vesical wall. On section it exhibits a central pedicle covered with long, closely packed, villous processes. Histologically the pedicle is composed of fibrous tissue and unstriped muscle, infiltrated with small inflammatory cells, while the villous processes resemble simple papillomata in their structure.

Isabel R., æt. 55, was admitted under Mr. Golding Bird. Her illness began eighteen months before admission with pain and severe hæmorrhage from the bladder. The tumour was removed by suprapubic cystotomy, and death occurred on the fifth day after the operation from cellulitis. *See Insp.* 1888, No. 47 ; and *Brit. Med. Jour.* 1889, vol. i. p. 17.

1793   Villous Growths in the Bladder.

The neck and posterior wall of a bladder mounted to shew upon the mucous membrane of the right side two villous tumours, one sessile and the other pedunculated, the latter, which is the larger, being of about the size of a cherry. At the neck of the bladder and in the prostatic urethra are several similar smaller growths. Histologically there is no evidence of malignant disease.

Richard W., æt. 42, was admitted under Mr. Golding Bird with a malignant growth in the floor of the mouth and died eight days after an operation for its removal. At the autopsy the lower lobe of the left lung was found to be affected by septic bronchopneumonia, and there were three recent ulcers in the duodenum. *See Insp.* 1894, No. 292.

1794   Epithelioma of the Bladder.

A bladder laid open to shew upon its left side between the ureter and the neck of the organ an oval plaque of growth measuring two inches in its long diameter. The growth has raised, somewhat everted edges, and its surface is shaggy. Histologically it has the structure of a squamous-celled epithelioma.

1795   Epithelioma of the Bladder.

A bladder laid open in front; the cervix of the uterus is attached below.   The whole of the interior of the bladder, except a small triangular area on its posterior wall, is covered with a shaggy new growth, which has destroyed the mucous and submucous coats, and in some places has invaded the surrounding fatty tissue. Histologically the growth is a squamous-celled epithelioma.   *Presented by T. Callaway, Esq.*

1796   Epithelioma of the Bladder.

A bladder laid open from behind to shew its wall extensively infiltrated by malignant deposit. At the apex the coats of the organ are greatly thickened and the serous surface presents several nodular excrescences. The mucous membrane of the anterior wall is marked by nodules and ridges which are coated with phosphatic deposit.   Histological examination of these portions of the wall shews the existence in the submucous tissue of a growth having the characters of a squamous-celled epithelioma.

1797   Epithelioma of the Bladder.

A portion of a bladder, the posterior wall of which is covered by a flattened mass of new growth. It occupies an area about three inches in diameter, and is composed of polvpi closely packed together, so that the surface resembles a cauliflower.   Around the chief growth are buds of various sizes. On section the growth is seen to blend with the muscular coat of the bladder.   The ureters are somewhat dilated.   Histologically the tumour is composed of large epithelial cells of squamous type arranged in columns or in alveolar spaces.

James C., æt. 50, was admitted in 1831 for hæmaturia of several months' duration.   At the autopsy there were secondary deposits in the liver, and the kidneys were hydronephrotic. *See Insp.* vol. 10, p. 154.

1798   **Epithelioma of the Bladder.**

A bladder opened in front, the wall of which is everywhere much thickened by new growth, except a small
portion of the anterior surface. The interior of the
bladder shews nodules of malignant deposit and extensive
ulceration. On the right side the growth has penetrated
the wall of the bladder and forms a large mass externally.
Histologically the growth is composed of narrow alveoli
filled with large squamous cells. No cell-nests were
seen.

> Francis D., æt. 52, was admitted under Mr. Cock for urinary
> symptoms and œdema of the right leg. At the autopsy there
> was a large mass of growth in the pelvic cavity, extending into
> the right iliac fossa, and obstructing the iliac vein. The liver
> contained a few small secondary nodules. *See Insp.* 1870, No. 75.

1799   **Epithelioma of the Bladder.**

A bladder opened in front, shewing on its posterior
wall a flat ulcerated growth, with thick, raised, everted
edges. Histologically it is a squamous-celled epithelioma with numerous cell-nests.

> Henry A., æt. 69, was admitted under Mr. Cooper Forster for
> urinary symptoms. Perinæal section had been performed thirty
> years previously for impermeable stricture, and the patient had
> been in the habit of passing a catheter through the fistula since
> the operation. It was supposed that the growth was the result
> of the constant irritation by the point of the catheter. *See Insp.*
> 1875, No. 246 ; and *Trans. Path. Soc.* vol. 28, p. 167.

1800   **Epithelioma of the Bladder.**

A bladder opened in front to shew extensive infiltration
of its wall by a soft new growth. Anteriorly the coats
of the organ are much thickened, and everywhere the
mucous membrane is destroyed by ulceration. At the
trigone is seen an incision left after lithotomy. The
ureters are considerably dilated. Histologically the
growth is a squamous-celled epithelioma with cell-nests.

Alfred W., æt. 28, was admitted under Mr. Lucas for symptoms of stone in the bladder of two years' duration. A large calculus was removed by lithotomy a fortnight before death. At the autopsy there were secondary deposits in the lumbar glands, and abscesses in the kidneys. *See Insp.* 1878, No. 196.

1801   **Epithelioma of the Bladder.**

The left half of a bladder, the cavity of which is nearly filled with a soft spongy new growth, which springs from a small area of the posterior wall, and had a wide attachment to the right lateral wall. The remainder of the vesical wall, which is closely opposed to the surface of the tumour, is unaffected by growth. Histologically it is a squamous-celled epithelioma.

William G., æt. 46, was admitted under Mr. Durham with hæmaturia and symptoms of cystitis. His illness began eighteen months before death with severe hæmorrhage from the urethra. Perineal section was performed, and four ounces of soft growth removed from the bladder. At the autopsy there was suppurative nephritis, and secondary deposits were found in the lumbar glands. *See Insp.* 1884, No. 154; and *R. C. S. Catalogue,* No. 3704 A.

1802   **Epitheliomatous Tumour of the Bladder.**

A bladder laid open by a frontal section to shew a globular mass of growth, two and a half inches in diameter, attached by a broad base to the posterior and left lateral walls of the organ. The surface of the growth is ulcerated, and the cut section shews it to have destroyed the muscular coats of the organ. The opening of the left ureter is immediately in front of the tumour. Histologically the growth has the structure of a squamous-celled epithelioma.

John M., æt. 69, was admitted under Dr. Washbourn for abdominal pain and constipation associated with a tense swelling in the left lumbar and inguinal regions. Some years previously he had suffered from hæmaturia and was sounded for stone. On admission the urine contained pus. The inguinal swelling was incised, and two pints of blood-stained fluid containing pus and urea were withdrawn. He died nine days after the operation and at the autopsy a perinephric abscess was found communicating with the above-mentioned cavity. The lower portion of the left ureter was dilated and its mucous membrane inflamed. The bladder contained two calculi, and presented evidences of old and recent inflammation. *See Insp.* 1893, No. 53.

1803   Epithelioma of a Sacculated Bladder.

A bladder with a somewhat enlarged prostate opened from the front, and shewing on its left side a sacculus measuring five and a half inches in its longest diameter. The neck of the sacculus, which is nearly three inches across, is infiltrated and greatly thickened by a new growth having a shaggy and villous surface. The growth extends to the anterior wall of the organ and occludes the left ureter. On the reverse of the specimen the left ureter, vas deferens, and vesicula seminalis are seen to lie upon the posterior wall of the cyst. Histologically the growth has the characters of a squamous-celled epithelioma, and the wall of the sacculus is composed of fibrous tissue and has no lining of mucous membrane.

John T., æt. 63, was admitted under Mr. Lucas for hæmaturia, from which he had suffered at intervals for the preceding four months. On admission a swelling was noticed above the pubes to the left of the middle line, extending upwards halfway to the umbilicus. The patient died suddenly four days later, and at the autopsy the right kidney was found to be enlarged and suppurating, the left being hydronephrotic. There were no secondary deposits. *See Insp.* 1891, No. 474; and *Trans. Path. Soc.* 1896, p. 155.

1804   Carcinoma of the Bladder.

A bladder opened in front, presenting a pedunculated tumour, the size of a walnut, situated around the orifice of the right ureter. A vertical section shews that it is composed of a short thick pedicle covered with long, closely-set, villous processes. The mucous membrane in the neighbourhood of the tumour and around the orifice of the left ureter shews small, slightly-raised nodules of similar growth. Histological examination shews that the papillary processes are attached to a pedicle composed of fibrous tissue, and that the muscular coat is infiltrated with columns of epithelial cells.

From a patient who had suffered from hæmaturia and symptoms of vesical calculus.

1805   Carcinoma of the Bladder.

A portion of a bladder shewing on its posterior wall, just above the orifice of the left ureter, a rounded, sessile tumour resembling a button-mushroom, and measuring one inch across. On section the base of the growth is seen to be continuous with the deeper structures of the vesical wall, and its surface to be covered with closely-set villous processes a quarter of an inch in length. Histologically the base of the tumour shews alveolar spaces filled with large spheroidal epithelial cells, and the villous processes have the structure of papillomata.

Joseph G., æt. 45, was admitted under Mr. Key in 1826 for irritability of the bladder and frequent micturition. A few years previously he had been successfully cut for stone. At the autopsy the kidneys were found to contain numerous abscesses, and there was suppurative peritonitis. *See Insp.* vol. 1, p. 59.

1806   Carcinoma of the Bladder.   Perforation.

The female genital organs with the bladder, which has been laid open to shew upon its posterior wall a mass of growth about an inch in diameter, having a ragged and ulcerated surface. At the upper part of the growth there is a perforation large enough to admit the tip of the little finger. The walls of the viscus are thickened, and its peritoneal coat nodulated by a malignant deposit having the characters of a spheroidal-celled carcinoma. The uterus is prolapsed, and the right ureter is dilated so as to measure rather more than an inch in diameter.

Faith D., æt. 46, was admitted under Mr. Key in 1833, and died from peritonitis. At the autopsy the right kidney was found to be hydronephrotic. *See Insp.* vol. 17, p. 97; and *Prep.* 1708.

1807 Carcinoma of the Bladder.

A bladder opened in front shewing two sessile papillary tumours near the orifice of the right ureter. The surface of the larger growth is covered with adherent blood-clot. Histological examination shews that they are composed of closely-packed villous processes, springing from the mucous membrane, and that the subjacent coats of the bladder are invaded with spheroidal-celled carcinoma. At the bottom of the preparation-jar are seen portions of laminated blood-clot found in the bladder.

John F., æt. 65, was admitted under Dr. Addison in 1842, and died suddenly in the taking-in room. He had previously been under treatment for stricture of the urethra. At the autopsy there was œdema of the larynx and feet, the kidneys were small and granular. *See Insp.* vol. 31, p. 259.

1808 Carcinoma of the Bladder.

A portion of the right half of a bladder, which shews a flat ulcerated growth occupying the mucous membrane above the orifice of the right ureter. The cut surface shews that the malignant disease has invaded the muscular coat of the organ. Histologically the growth is a spheroidal-celled carcinoma with abundant stroma.

From a gentleman, aged 50, in whose urine Dr. Rees detected " cancer cells."

Presented by Mr. Roper.

1809 Carcinoma of the Bladder.

A bladder opened from behind and everted to shew its anterior wall covered by a shaggy new growth arranged in three large patches, with healthy mucous membrane intervening. The surface of the growth is ragged from ulceration, and its base extends into the deeper layers of the vesical wall. Histologically the growth consists of squamous and spheroidal cells enclosed in large alveoli with scanty stroma.

Thomas T., æt. 41, was admitted under Mr. Callaway. He had suffered all his life from symptoms of stone, and since the age of twenty-one from occasional hæmaturia. At the autopsy a large oxalate calculus was found in the bladder, and there were secondary deposits of carcinoma in the lumbar glands. *See Insp.* 1854, No. 136.

1810   Carcinoma of the Bladder.

A bladder opened in front to shew its apex and a portion of the posterior wall thickened by a new growth. The deposit is situated in the muscular coat, the mucous membrane and peritoneum being unaffected. Histologically the growth is a spheroidal-celled carcinoma.

Susan G., æt. 56, was admitted under Dr. Gull for numbness and loss of power in the right arm. The right breast was affected with carcinoma. At the autopsy secondary deposits were found in the lungs, pleura, and mediastinal glands, and there was a large mass of growth in the right axilla invading the brachial plexus. *See Insp.* 1855, No. 36; *Prep.* 1620 (15) [2nd Edit.]; *Drawing*, 88 (51); and *Trans. Path. Soc.* vol. 41, p. 180.

1811   Carcinoma of the Bladder.

A bladder shewing its posterior and left lateral walls occupied by a new growth, which involves the whole thickness of the vesical wall, and forms a prominent rounded tumour with an extensively ulcerated surface. On the reverse of the specimen nodular masses of malignant deposit are seen to project from the outer surface of the viscus, and one of such nodules blocks the canal of the left ureter. Histologically the growth is a spheroidal-celled carcinoma with scanty stroma.

James B., æt. 62, was admitted under Mr. Cock with cystitis and hæmaturia. He had suffered for two years from urinary symptoms, and had sometimes passed small calculi. At the autopsy there was acute suppurative nephritis, but no secondary deposits of growth were found in other organs. *See Insp.* 1855, No. 200.

1812 Carcinoma of the Bladder.

A bladder laid open to shew a rounded mass of growth, one inch and a half in diameter, which projects from the posterior wall around the orifice of the right ureter. The section shews that there is a broad pedicle extending into the muscular coat, and that the surface is composed of closely-packed villous processes. A small flattened mass of apparently similar structure is seen to the left side of the main tumour. Histologically the growth is a carcinoma with large alveolar spaces filled with spheroidal cells and lined with a layer of columnar epithelium.

Joseph B., æt. 55, was admitted under Dr. Gull for hæmaturia of eight months' duration, having suffered from incontinence of urine for twelve months. He died from loss of blood six weeks after admission, and at the autopsy the pelves and infundibula of the kidneys were found to be much dilated. *See Insp.* 1860, No. 149.

1813 Carcinoma of the Bladder.

A bladder laid open in front to shew upon its posterior wall a flat mass of new growth, measuring between three and four inches in diameter. In the centre of this area there is a dark excavation surrounded by numerous flattened nodules. Histologically the growth is a carcinoma, the alveoli of which are large and filled with spheroidal cells.

From an aged female who had long suffered from hæmorrhage from the bladder.

Presented by Mr. Lacey, 1864.

1814 Carcinoma of the Bladder.

A portion of a bladder shewing around the orifice of the left ureter a lobulated new growth an inch and a half in diameter. The tumour is raised half an inch above the level of the mucous membrane, and its surface has a cauliflower-like appearance. The vertical section shews that the growth is composed of closely-packed villous processes attached to a broad pedicle which invades the muscular coat of the bladder. Histologically the portion of the growth within the muscular coat presents large alveolar spaces lined with two or three layers of columnar epithelium and filled with large spheroidal cells.

Francis S., æt. 40, was admitted under Mr. Davies-Colley for hæmaturia and cystitis. There was a history of pain on micturition and the occasional passage of blood in the urine during the three years before admission. At the autopsy there was suppurative pyelo-nephritis with chronic tuberculosis of the lungs. *See Insp.* 1889, No. 399.

1815   **Carcinoma of the Bladder.**

A bladder mounted to shew a uniform infiltration of its walls with malignant growth. The cut section shews that the wall measures as much as an inch in thickness, and that its coats are replaced by a firm white deposit. The interior of the organ is shaggy and ulcerated, and exteriorly the growth forms a large nodular excrescence projecting from the apex on the left side. Histologically the growth has the character of a carcinoma with scanty stroma, most of the epithelial cells being of spheroidal type, whilst a few are large and squamous.

John J., æt. 48, was admitted under Mr. Lucas passing urine which contained pus and blood. Seven months previously an exploratory operation shewed that his bladder was infiltrated with malignant growth. He died nine days after admission, and at the autopsy both kidneys were found to contain abscesses, the right being very small and fibrous. *See Insp.* 1893, No. 233.

1816   **Carcinoma of the Bladder. Perforation.**

A bladder laid open in front, the right wall of which is occupied by a soft new growth. The centre of the growth is deeply ulcerated, and communicates by a ragged opening with the peritoneal cavity. The bladder is much hypertrophied, and the right ureter is dilated. Histologically the wall of the bladder is infiltrated with narrow bands of spheroidal epithelial cells embedded in a fibrous stroma.

1817   Villous Carcinoma of the Bladder.

A bladder laid open from behind to shew a large mass of new growth attached close to the orifice of the left ureter, and covered by long delicate villous processes. The growth is hard and invades the muscular coat. There are tufts of simple papillomata on the surrounding mucous membrane, and at the apex of the organ there is an isolated rounded area of new growth bearing very short villi. The left ureter, marked by a red rod, is dilated. Histologically the growth is a spheroidal-celled carcinoma with abundant stroma.

Presented by Dr. Rees, 1856.

1818   Sarcoma of the Bladder.

A bladder laid open to shew a polypoid tumour attached to its neck on the left side and measuring two and a half inches in length. A vertical section of the polypus shews that it is situated upon a flattened disc of growth in the submucous tissue surrounding the neck of the bladder. Histologically the tumour is composed of round and spindle-shaped sarcomatous cells, and the surface is covered with stratified epithelium. In some parts the cells have undergone myxomatous degeneration.

Thomas W., æt. 18 months, was admitted under Mr. Birkett with symptoms of stone in the bladder, of some months' duration, and died a week after admission. At the autopsy acute suppurative nephritis was discovered. *See Drawing,* 309 (20); and *Insp.* 1865, No. 44.

1819   Sarcoma of the Bladder.

A bladder opened in front to shew a lobulated sessile tumour, the size of a walnut, attached to the right lateral wall an inch from the orifice of the ureter. At its periphery the growth is smooth and covered by healthy mucous membrane, but in the centre the surface is shreddy from ulceration. Histologically the tumour is a spindle-celled sarcoma, and does not invade the muscular coat.

Joseph V., æt. 34, was admitted under Dr. Habershon for recurrent hæmaturia, which was first noticed at the age of twelve years. He died from anæmia. At the autopsy no growths were found in any other organ. *See Insp.* 1877, No. 189.

1820   Sarcoma of the Bladder.

The left half of a bladder divided by sagittal section to shew its cavity entirely filled with malignant deposit. The organ measures five inches in its longest diameter, and its coats, except for a small area at its base, are entirely replaced by a firm homogeneous growth. Externally the growth forms a prominent rounded mass, and it has the histological characters of a small round-celled sarcoma.

From a man, æt. 68, who had suffered for two and a half years from pain in the loins and occasional hæmaturia. At the autopsy the ureter and pelves of the kidneys were found to be dilated. *See Trans. Path. Soc.* 1885, p. 284.

Presented by Mr. Farrant Fry.

1821   Sarcoma of the Bladder and Prostate.

A bladder opened in front, the interior of which shews a pedunculated tumour attached to the wall near the orifice of the right ureter. The surface of the tumour is covered with numerous pyriform polypi, and the internal meatus of the urethra is thickly beset with similar growths. Below the base of the bladder and tho prostatic urethra there is a large oval tumour measuring three inches in its longest diameter. On the reverse of the specimen a vertical section of this tumour shews that it has a gelatinous appearance, and that there is a central cavity due to degenerative changes. Histologically the growths have the characters of a sarcoma composed of round and oval cells.

Richard T., æt. 2, was admitted under Mr. Hilton for a tumour of the bladder which produced a visible swelling above the pubes and was also felt per rectum. Death was due to suppurative nephritis and peritonitis. *See Insp.* 1858, No. 169.

1822　**Sarcomatous Polypi of the Bladder.**

A bladder considerably dilated and laid open to shew its cavity filled with large polypoid masses of new growth. They measure three or four inches in length, and have a broad attachment two inches in diameter to the apex and right lateral wall. The surface of the growth is ulcerated, except at its base, where it is covered by healthy mucous membrane. The remainder of the interior of the bladder is unaffected. Histologically the growth is a spindle-celled sarcoma.

1823　**Sarcomatous Polypi of the Bladder.**

A portion of a bladder everted to shew numerous soft polypi hanging in clusters from its base and posterior wall. They are angular in shape, and some of them have very thin thread-like pedicles. In the centre of the largest group is seen the orifice of the right ureter. The mucous membrane surrounding the attachments of the polypi is remarkably thickened and ridged, and has a very definite line of demarcation from the healthy mucous membrane at the apex. On the reverse of the specimen are seen two large cysts which have been laid open. The one on the right is a dilatation of the lower end of the right ureter, the undilated portions of which, above and below the cyst, are indicated by blue rods.

P 2

The cyst on the left is unconnected with the corresponding ureter, and has a single opening into the prostatic
urethra, marked by a red rod.   It appears to be due to
a dilatation of the prostatic glands.   Histologically the
polypi, as well as the thickened mucous membrane, are
composed of oval and spindle-shaped cells, the latter
arranged in interlacing bundles.   They are entirely
limited to the submucous coat, and have a normal epithelial covering.

From a boy, about 14 years of age, whose symptoms resembled
those of stone in the bladder.   *See Drawing*, 369 (5).

Presented by Sir Astley Cooper.

1824   Sarcomatous Polypi of the Bladder.

A female bladder laid open from behind to shew a mass
of polypi attached to its neck and anterior wall.   Some
of them are pear-shaped and have short pedicles, while
others are sessile and are warty on the surface.   A
portion of the growth protrudes through the meatus
urinarius and forms a globular tumour externally.
Histologically the growth is a myxo-sarcoma.

Sarah J., æt. 5, was admitted under Mr. Birkett for urinary
symptoms of two months' duration.   A tumour protruded through
the meatus of the urethra and was removed by ligature.   She
died of suppurative nephritis one month after admission.   *See
Insp.* 1858, No. 10; *Drawings,* 369 (10 & 11) ; and *Trans. Med.-Chi.
Soc.* vol. 41.

1825   Bladder invaded by Sarcoma.

A portion of a bladder mounted to shew around the
situation of the orifice of the right ureter a flat mass of
growth measuring two inches in its longest diameter,
and projecting about half an inch above the surrounding
mucous membrane.   The surface of the growth presents
a lobulated appearance, and is in parts superficially
ulcerated.   On the reverse of the specimen a transverse
section through the ureter shews it to be enlarged

and completely occluded by a white mass of growth which histologically has the characters of a small round-celled sarcoma.

Edwin O., æt. 54, was admitted under Dr. Pitt with pains in the back, frequency of micturition, and occasional hæmaturia. A nodular tumour was felt in the right lumbar region, and the urine contained pus and blood-stained débris. The patient died twelve days after admission, and at the autopsy the right kidney was found to be infiltrated by a soft white growth, which invaded the pancreas and the duodenum, and extended down the ureter to the bladder. *See Insp.* 1807, No. 28.

1826   Bladder invaded by Sarcoma.

A sagittal section through a bladder and rectum shewing the former embedded in a mass of firm growth, which in the recent state filled the pelvic cavity. The malignant deposit is seen to infiltrate the muscular coat of the organ, and to appear in the form of flattened nodules beneath the mucous membrane, which is free from ulceration. Histologically the growth is a small round-celled sarcoma containing numerous hæmorrhages.

Robert R., æt. 27, was admitted under Mr. Howse with symptoms of paraplegia and intestinal obstruction. A pelvic growth was detected which gradually extended into the abdomen. The intestinal symptoms were relieved by colotomy, and the patient died about a year from the onset of the disease. At the autopsy secondary deposits were found in the skull, lungs, liver, iliac veins, and peritoneum. There was ascending pyelo-nephritis. It was thought that the primary seat of growth was in the pelvic lymphatic glands. *See Insp.* 1892, No. 357.

1827   Melanotic Sarcoma of the Bladder.

A bladder opened in front to shew a small polypoid tumour attached to the lateral wall about an inch from the orifice of the left ureter. It has a long slender pedicle and the surface is marked with patches of brown pigmentation. Histologically it is composed of round sarcomatous cells with granules of pigment scattered among them.

George C., æt. 32, was admitted under Mr. France for a melanotic tumour of the left eye, which had existed two and a half years. About a year before his death small black nodules appeared on the surface of the body. At the autopsy secondary deposits were found in the bones, liver, lungs, and other organs. *See Insp.* 1859, No. 119 ; *Preps.* 1400 (15), 1060 (60), and 2354 (20) [2nd Edit.].

1828   Hydatid of the Bladder.

A bladder with the lower end of the rectum mounted to shew between them a pyriform cyst measuring seven inches in its transverse diameter. The ureters, which are dilated, are seen to cross the sac and enter the bladder at right angles midway between the summit and the elongated neck of the viscus. The vesiculæ seminales lie upon and project on either side of the prostate, which is flattened by the pressure of the tumour. The muscular coats of the bladder and rectum are considerably hypertrophied. The cyst appears to have originated beneath the serous coat of the bladder, the upper part of the sac being covered by peritoneum. At the bottom of the preparation-jar are some daughter-cysts found in the sac.

James P., æt. 40, was admitted under Mr. Morgan in 1835 for incontinence of urine associated with a tumour in the hypogastric region. A catheter was passed into the bladder, and a few drachms of healthy urine were withdrawn. Subsequently an instrument was again passed, and a quantity of hydatid cysts and débris was evacuated. At the autopsy it was found that the catheter had perforated the urethra and entered the sac. The kidneys were extremely hydronephrotic. *See Insp.* vol. 12, p. 87 ; and *Guy's Hosp. Reps.* 1837, p. 464.

1829   Hydatid of the Bladder.

A bladder opened from the front and shewing attached to its posterior wall a portion of a cyst which in the recent state measured eight inches in diameter, and was filled with caseous and cretaceous material. The wall of the cyst is denso and fibrous, and is lined with rough calcareous plates.

Richard R., æt. 60, was admitted under Dr. Barlow for heart disease, from which, ten weeks later, he died. *See Insp.* 1854, No. 194.

1830   Hydatid of the Bladder.

The posterior wall of a bladder to which is attached a thin-walled pyriform cyst measuring seven inches from above downwards and five inches transversely. In the recent state it contained three pints of small hydatids. The ureters and vasa deferentia are adherent to the cyst-wall, and the vesiculæ seminales are pushed downwards from their normal position, the right one being bent backwards so as to lie on the posterior surface of the prostate.

William A., æt. 52, was admitted under Dr. Habershon for cancer of the stomach. He had also a tumour at the lower part of the abdomen corresponding in size and shape to a distended bladder. It could not, however, be emptied by the catheter. He suffered no inconvenience from the presence of the tumour, dying twenty-four days after admission from the disease for which he sought admission. At the autopsy the kidneys were found to be small and healthy. *See Insp.* 1860, No. 80.

1831   Hydatid of the Bladder.

The male pelvic viscera mounted to shew a large hydatid cyst situated between the bladder and the rectum. The cyst is eight inches in vertical measurement and by its extension downwards causes the lower end of the rectum to be separated from the prostate by an interval of four inches. On the right lateral wall of the bladder is seen the orifice of a small sacculus.

Thomas R., æt. 66, was admitted under Dr. Perry with symptoms of hydatids in the liver and peritoneal cavity. He had been in Australia, and for the seventeen years preceding admission had frequently been under treatment for hydatid disease. An incision was made into the sac in the liver, and three weeks later the patient died. At the autopsy hydatids were found in the left

lung and in the upper part of the abdominal cavity.   *See Insp.*
1895, No. 151.

1832   Hydatid of the Bladder.

A large number of daughter-cysts of various sizes which
were contained in the hydatid sac which forms the
preparation No. 1830.

1833   Hydatids passed in Micturition.

Numerous hydatid cysts varying in size from a line to
three-quarters of an inch in diameter which were voided
with the urine.

> Henry B., æt. 34, was admitted under Mr. Birkett in 1851
> bringing with him the hydatids which form the preparation, and
> which were stated to have been passed from the urethra during
> micturition. Ten years before admission he had suffered from pain
> in the left loin, and a year later noticed a swelling in that region.
> Ten months after the swelling appeared he passed " bladders and
> skins," and continued to pass similar bodies at varying intervals
> until the time of admission. While under observation no tumour
> could be detected. It was thought that the parent cyst was
> " seated near the left kidney, having a communication with the
> pelvis of this organ or its ureter." *See Guy's Hosp. Reps.* 1851,
> p. 300.

1834   Dermoid Tumours of the Bladder.

Portions of two tumours removed from the bladder by
operation. The lowest specimen is a rounded mass,
the surface of which is wrinkled and bears short hairs
covered with phosphatic deposit. The middle specimen,
to which a catgut ligature is attached, is the pedicle of
the tumour mounted below it. This has likewise a
thick coating of phosphates and hair. The upper specimen is a polypoid tumour the size of a filbert, having a
short thick pedicle and a tuft of hairs on the surface.
Histologically the tumours are covered with true skin
containing sebaceous and sweat glands, and the deeper
parts are composed of fibrous tissue.

From a married lady, æt. 30, who passed per urethram phosphatic calculi formed on long hairs. An exploratory operation was performed, and the lowest specimen in the preparation was removed. Ten days later the pedicle and ligature came away with the urine. As the urinary symptoms recurred a second operation was performed three weeks after the first, and the uppermost specimen was removed. Eight years after this operation the patient was in good health. *See Prep.* 2023.

Presented by Mr. Thomas Bryant.

1835   **Fatty Contents of a Dermoid Cyst of the Bladder.**

A quantity of fatty material which formed part of the contents of a cyst in the neighbourhood of the bladder.

From George M., æt. 21, who had occasionally suffered for some months from retention of urine. Subsequently a fluctuating tumour was discovered in the pelvis at the side of the bladder. The tumour was tapped per rectum and two pints of fluid fat were withdrawn which consolidated on cooling. The patient made a good recovery. *See Trans. Path. Soc.* vol. 13, p. 148.

Presented by Mr. G. W. Pretty, 1861.

1836   **Bladder perforated by Pelvic Abscess.**

A portion of a bladder mounted to shew on its posterior wall, a little behind the orifice of the right ureter, a circular opening measuring about one-eighth of an inch in diameter and communicating with an abscess cavity situated between the neck of the bladder and the lower end of the rectum. The walls of the viscus are hypertrophied and measure half an inch in thickness.

John R., æt. 45, was admitted under Dr. Bright in 1838, having suffered from occasional hæmaturia during the preceding five years. Some months before admission he noticed that he passed urine with his motions, and when in the hospital it was found that fæces were discharged per urethram and urine per anum. He died two days after admission, and at the autopsy a cancerous ulcer was found in the colon, which had established a communication with the summit of the bladder. The abscess cavity behind the bladder contained more than a pint of fæculent pus. *See Insp.* vol. 27, p. 75.

1837   **Post-vesical Abscess communicating with the Rectum.**

A bladder and the lower end of the rectum mounted to shew between them and communicating with each of them an abscess cavity, which in the recent state contained a considerable amount of pus. The wall of the bladder is somewhat thickened and its mucous membrane is partially destroyed by ulceration. On the posterior wall there are several openings of communication with the abscess, and in the prostatic urethra is a perforation also leading into this cavity.

John C. was admitted under Mr. Aston Key in 1826 for fractured spine. Paraplegia supervened and the patient died two months after the accident, a catheter having been occasionally passed on account of retention of urine. At the autopsy the left kidney was found in a condition of suppurative nephritis. *See Insp.* vol. 1, p. 17 ; and *Prep.* 1035 [2nd Edit.].

1838   **Vesical Fistula opening externally.**

A bladder, the summit and posterior wall of which have been removed to shew upon its anterior wall the opening of a sinus which leads through the horizontal ramus of the right pubic bone, and in the recent state opened externally over the centre of Poupart's ligament on the left side. The bone adjacent to the sinus is thickened and partially necrosed.

From a youth, æt. 18, who two years before his death noticed a swelling over the pubis, and six months later suffered from an abscess which discharged itself in the left groin. Six months after the abscess had opened he was a patient in the hospital, when pus was found in his urine and dead bone was felt with a probe in the neighbourhood of the symphisis pubis.

Presented by Mr. Cooper Forster, 1862.

1839   **Vesico-vaginal Fistula.**

The female genito-urinary organs mounted to shew a a large vesico-vaginal fistula. The fistula is situated at the neck of the bladder, is roughly triangular in shape, and readily admits two fingers. The wall of the bladder is somewhat hypertrophied.

From a woman, from whose vagina a calculus was removed having the size and form of a duck's egg.

Presented by Mr. Tipple.

1840 Prolapse of Ureter.

The urinary organs of a female child mounted to shew around the orifice of the left ureter a raised edge, external to which is a ragged border apparently consisting of villous growth. The ureter is much dilated and tortuous, and about half an inch from its distal extremity presents an imperfect septum. The kidney is hydronephrotic, and in the recent state was in a condition of suppurative nephritis. The right kidney and its ureter are normal.

From a female child, æt. 18 months, who presented a funnel-shaped protrusion, about an inch long and half an inch thick, at the meatus urinarius. A catheter introduced through an orifice at the apex of the funnel passed five inches towards the left side of the abdominal cavity, and gave exit to 4 or 5 ounces of fetid pus. The protruding mass was removed by Mr. Davies-Colley, and nine days later the child died. It was thought that the protrusion consisted of a prolapse of the mucous membrane of the ureter. *See Trans. Path. Soc.* vol. 30, p. 310.

1841 Calculus at the Orifice of the Ureter.

A portion of the bladder with the lower half of the right ureter. The orifice of the ureter is blocked by a large irregular calculus measuring rather more than half an inch in diameter, the greater part of which protrudes into the cavity of the bladder. A chain of similar calculi extends up the ureter, the lumen of which is dilated and its walls markedly hypertrophied. The left ureter is much shrunken and is of smaller size than the corresponding vas deferens.

Presented by Mr. J. Parrot.

1842   Encysted Calculus in the Bladder.

A bladder laid open to shew in its right lateral wall by
the side of the ureter a sacculus in which is contained
a rough pear-shaped calculus measuring an inch in
length.   The right ureter appears to be compressed
by the sacculus and its contents, whilst above the
obstruction it was dilated.   In the urethra is seen
the opening left after lithotomy.

David D., æt. 38, was admitted under Mr. Birkett, who
removed a calculus from his bladder by perinæal lithotomy. Five
days later the patient died, and at the autopsy the left kidney
was found to be hydronephrotic, and the right in a condition of
pyelonephritis. *See Insp.* 1860, No. 116.

1843   Calculus in the Bladder.

A bladder laid open to shew its cavity almost completely filled by a white phosphatic calculus measuring
three inches from apex to base.   The anterior surface of
the calculus is rough and flattened, whilst posteriorly
it is smooth and moulded to the shape of the cavity.
The ureters are considerably dilated and the walls of
the bladder are thickened.

George H., æt. 52, was admitted under Mr. Bransby Cooper,
in 1829, for calculus of the bladder. No operation was attempted,
and about three weeks after admission the patient died. At the
autopsy the kidneys were found to be hydronephrotic and to
contain numerous small abscesses. *See Insp.* vol. 14, p. 104.

1844   Calculus in the Bladder.

A bladder considerably hypertrophied and shewing an
oval calculus about two inches in diameter firmly impacted at its summit.   The prostate is greatly enlarged
and its middle lobe is very prominent.

1845   Calculus in the Bladder.   Operation.

A bladder, the cavity of which is considerably contracted and its walls thickened.   It has been laid open

to shew a nodular calculus three-quarters of an inch in diameter lodged at its neck. The prostate is greatly enlarged.

William H., æt. 70, was admitted under Mr. Bryant with symptoms of stone in the bladder, for the relief of which lateral lithotomy was attempted. The stone could not be removed and the patient died on the day following the operation. At the autopsy the right kidney was found to be contracted and to contain a calculus; the left exhibited compensatory hypertrophy. *See Insp.* 1862, No. 198.

1846   Concretions in the Bladder.

A child's bladder, the walls of which are much hypertrophied. The proximal portion of the urethra is dilated, and its mucous membrane is deeply ulcerated. Below are mounted two ovoid concretions, one slightly larger than the other and measuring an inch and a quarter in its longest diameter. On section the bodies present concentric translucent laminæ separated from each other by a dull white phosphatic material. The external surface is sanded over with a similar deposit.

From a male child, æt. 2, who suffered from symptoms of stone in the bladder. He was sounded, but no stone was detected. *See Guy's Hosp. Reps.* 1837, p. 268; and *Drawing*, 369 (75).

1847   False Membrane after Lithotomy.

A portion of tough fibrinous membrane coated with phosphatic deposit which was removed from the perinæal wound after lateral lithotomy.

Presented by Mr. Hilton, 1850.

1848   Lateral Lithotomy.   Partial Removal of the Prostate.

A small piece of a prostate gland which is stated to have come away during the performance of lateral lithotomy. With it is mounted the half of the calculus.

James P., æt. 67, was admitted under Mr. Bryant, in 1875, with vesical symptoms of eighteen months' duration. *See Surgical Reps.* 1875, Case 243.

1849   Lateral Lithotomy with partial Prostatectomy.

A portion of the prostate gland which was removed during the operation of lateral lithotomy. With it is mounted a half of the calculus on account of which the operation was performed.

From Thomas M., æt. 70, who had suffered from symptoms of stone in the bladder for four years, for the relief of which lithotomy was performed by Mr. Bryant. The patient made a good recovery and was well six months after the operation. *See Surgical Reps.* 1876, Case 255.

1850   Bladder wounded in Lithotomy.

A child's bladder shewing upon its posterior wall, midway between its summit and neck a little to the right of the middle line, a punctured wound about an eighth of an inch in diameter. On the left side of the prostate is seen the incision of lateral lithotomy.

William C., æt. 5, was admitted under Mr. Aston Key, in 1845, with symptoms of vesical calculus, for which lithotomy was performed. The patient died twenty hours after the operation, and at the autopsy the peritoneum was found to be studded with tubercles and there was recent peritonitis. *See Insp.* vol. 34, p. 61.

1851   Peritoneum perforated in Lateral Lithotomy.

The pelvic viscera of a male child mounted to shew the condition of the parts after an unsuccessful operation of perinæal lithotomy. The track of the gorget is seen passing across the bladder and terminating in a large perforation at the bottom of the recto-vesical pouch.

1852   Fistulæ after Lithotomy.

A vertical section through the bladder and rectum shewing immediately in front of the prostate a sinus, indicated by a blue rod, which communicates with the cellular tissue in the ischio-rectal fossa.

Thomas G., æt. 18, was admitted under Mr. Davies-Colley with fistulæ in the groin, perinæum, and in front of the scrotum.

Seven years previously the patient had undergone lithotomy, the wound of which had never healed. Cock's operation was performed, and twelve days later the patient died with symptoms of peritonitis. At the autopsy a perforation was found in the bladder communicating with the peritoneal cavity, which was thought to have been produced by the catheter tied in after the operation. *See Insp.* 1881, No. 270; and *Prep.* 1757.

1853   Bladder Eleven Years after Lithotomy.

A bladder laid open to shew in the prostatic urethra a depressed scar, anterior to which is a short blind sinus. On the reverse of the specimen is seen in the perinæum the healed wound of lateral lithotomy.

Frederick G., æt. 14, was admitted under Dr. Habershon for acute rheumatism, and died from pericarditis and myocarditis one month after admission. Eleven years previously he was operated on for stone, and since that time had suffered from incontinence of urine. *See Insp.* 1876, No. 10; and *Trans. Path. Soc.* vol. 27, p. 208.

1854   Bladder Twenty-four Years after Lithotomy.

A bladder with the prostate and the root of the penis laid open to shew on the floor of the prostatic urethra an oval ulcer divided by a white fibrous band, and exposing in its base the subjacent muscular tissue. The caput gallinaginis is deflected to the right.

Joseph L., æt. 29, was admitted under Dr. Pavy for phthisis, from which some months later he died. Lithotomy was performed when the patient was five years of age. *See Insp.* 1874, No. 27; and *Trans. Path. Soc.* vol. 25, p. 186.

SECTION XXXIV. — CALCULI, CONCRETIONS, AND FOREIGN BODIES FROM THE URINARY BLADDER.

Uric Acid Calculi : 1855–1868.
Oxalate of Calcium Calculi : 1869–1878.
Urate of Ammonium Calculi : 1879.
Cystine Calculi : 1880–1883.
Xanthic Oxide Calculus : 1884.
Phosphate of Calcium Calculus : 1885.
Triple Phosphate Calculus : 1886.
Fusible Calculi : 1887–1891.
Phosphatic Calculi : 1892–1896.
Carbonate of Calcium Calculus : 1897.
Urate of Sodium Calculus : 1898.
Compound Calculi : 1899–1920.
Alternating Calculi : 1921–1936.
Mixed Calculi : 1937–1995.
Dumb-bell Calculus : 1996.
Miscellaneous Calculi : 1997–1999.
Cast of a Calculus : 2000.
Pretended Calculi : 2001.
Calculi passed and removed per urethram : 2002–2008.
Calculi passed and removed through fistulæ : 2009–2012.
Foreign Bodies from the Bladder : 2013–2022.

1855   **Uric Acid Calculus.**

The two halves of an ovoid calculus measuring 2¾ inches in its longest diameter and weighing 1531 grains. The surface is of a dark brown colour and is coarsely granular; its structure consists of uric acid disposed in concentric layers around a small central nucleus. It was analysed by Dr. B. Babington.

" This wonderful stone, by the blessing of God, was happily extracted from Mr. W. by the judicious Mr. Richard Lambert, of Newcastle, on the 17th of June, 1760, at the age of 56, who lived seven years after and worked at his trade."

1856   Uric Acid Calculus.

The half of a rounded calculus weighing 309 grains and found by Dr. Babington to be formed of uric acid. On section the outer laminæ are seen to be compact, while in the centre there is a cavity, the wall of which is composed of closely-set spicules of uric acid. The external surface is of a dark brown colour, smooth in parts and elsewhere finely granular.

1857   Uric Acid Calculus.

The two halves of a calculus which weighed 258 grains and was successfully removed by lithotomy. On section it is seen to be composed of concentric laminæ of a yellowish-white colour, and its surface, which is brown, is nodular and tuberculated. The calculus was analysed by Dr. Babington, and found to consist of pure uric acid.

From George V., æt. 14, a patient of Mr. Bransby Cooper.

1858   Uric Acid Calculi.

The halves of two calculi which weighed 207 grains and 178 grains respectively, and were removed by lithotomy. Their exterior is irregularly nodulated and covered with very minute crystals. Internally the compact and yellowish-white substance of which they are composed presents numerous fissures and excavations. The calculi were analysed by Dr. Babington and found to consist of uric acid.   *Presented by Sir Astley Cooper.*

1859   Uric Acid Calculi.

The halves of two uric acid calculi weighing 465 and 416 grains respectively.

From James P., æt. 65, from whom in 1851 the larger calculus was removed by Mr. Bransby Cooper during life, whilst the smaller one was found after death impacted in a sacculus of the bladder.

1860   **Uric Acid Calculus.**

An ovoid calculus measuring an inch and three quarters in its longest diameter, and divided to shew internally a laminated structure, whilst externally its surface is minutely granular and presents several small nodular excrescences. Its weight is 560 grains, and it consists of uric acid.

From Harry P., æt. 6, upon whom lithotomy was performed by Mr. Cock in 1859.

1861   **Uric Acid Urethral Calculus.**

An elongated calculus measuring half an inch in length and a quarter of an inch across, and composed of uric acid. It has a brown colour and a rough granular surface.

From a child, æt. 3½ years, a patient of Mr. Callaway, jun.

1862   **Uric Acid Urethral Calculus.**

The half of a calculus, resembling a date-stone in size and shape, which was removed from the urethra. It is composed of uric acid.

Presented by Mr. Cooper Forster.

1863   **Uric Acid Calculi.**

Three rounded calculi, each of them measuring about an inch and three quarters in diameter and having a smooth exterior. One of them has been divided, and is seen to consist of a compact grey nucleus surrounded by two zones of similar material, separated from each other and from the nucleus by a thin layer of a darker more friable substance. The outermost zone shews several radial fissures.

Removed, after death, from the bladder of an elderly man, who during life had suffered little inconvenience from their presence.

1864   Uric Acid Calculus.

A small uric acid calculus weighing 120 grains and presenting a laminated structure and an irregular dark brown surface.

Stephen P., æt. circa 50, was admitted under Mr. Bransby Cooper in 1828 with symptoms of stone in the bladder, for the relief of which lithotomy was performed. The patient died thirty-two hours after the operation. *See Insp.* vol. 5, p. 152.

1865   Uric Acid Calculus with two Nuclei.

The half of a uric acid calculus somewhat pyriform on section, and apparently produced by the coalescence of two smaller calculi formed around separate nuclei, one of which is spherical, whilst the other is lenticular.

1866   Uric Acid Calculus.

The half of an ovoid calculus measuring an inch and a quarter in its longest diameter, which was removed from the bladder by lithotomy. The nucleus and the outer well-defined concentric laminæ of the concretion are dense, whilst the intermediate zone is of open texture, the small interstices having a radial arrangement.   *Presented by Mr. Aston Key,* 1839

1867   Uric Acid Calculi passed by the Urethra.

A considerable number of spherical calculi varying in size from a sixteenth to a quarter of an inch in diameter, which were passed during micturition. They are of a light brown colour, have a smooth surface, and consist of uric acid.

1868   Uric Acid Calculus.   Lithotrity.

The fragments of a large uric acid calculus removed from the bladder in the operation of lithotrity.

Presented by Mr. Aston Key.

Q 2

1869   Oxalate of Calcium Calculus.

A mulberry calculus measuring an inch in diameter, the brown tuberculated surface of which is sprinkled with minute glistening crystals of oxalate of calcium.   It was analysed by Dr. Golding Bird.

From William R., æt. 14, upon whom Mr. Aston Key performed lithotomy in 1830.

1870   Oxalate of Calcium Calculus.

An oval calculus measuring an inch and three quarters in its longest diameter and having a brown nodular surface.   The cut section shews it to be composed of a pale brown material, the central portions having a worm-eaten appearance, whilst the outer zone is hard and compact.   The calculus was analysed by Dr. Rees and found to consist of oxalate of calcium.

From a patient upon whom Mr. Bransby Cooper operated successfully in 1837.

1871   Oxalate of Calcium Calculus.

A calculus of irregular shape measuring an inch and a half in its longest diameter and formed of a dense yellowish material, which was found on analysis by Dr. Golding Bird to be oxalate of calcium.   The exterior of the calculus is covered with short thick spinous processes.   *Presented by Mr. Aston Key.*

1872   Oxalate of Calcium Calculus.

An oval calculus measuring an inch in its longest diameter and composed of a dense substance arranged in wavy concentric layers.   The exterior of the calculus is brown, and its surface somewhat resembles that of a "hobnailed" liver.   It was analysed by Dr. Odling and found to consist of oxalate of calcium.

G B. æt. 12, was admitted under Mr. Cooper Forster in 1857 with symptoms of stone in the bladder of many years' duration, for the relief of which lithotomy was performed.

1873   Oxalate of Calcium Calculus.

The half of a rounded calculus two inches in diameter presenting a dark brown exterior, which is coarsely nodular and partially encrusted with phosphatic deposit. The cut surface is polished and is composed for the most part of a dark brown substance with lighter zones, the whole consisting of oxalate of calcium.

Presented by Mr. Callaway.

1874   Oxalate of Calcium Calculus.

The half of a rounded calculus, the cut surface of which shews a dense nucleus and body of a dark brown material surrounded by a zone of lighter colour. The exterior is tuberculated and has the appearance characteristic of a mulberry calculus. Its weight is 290 grains.

From a patient, æt. circa 45, upon whom Sir Benjamin Brodie performed lithotomy in 1842. Lithotrity was attempted and found impracticable on account of the hardness of the stone.

Presented by Mr. Bransby Cooper.

1875   Oxalate of Calcium Calculus.

A rough mulberry calculus, half an inch in diameter, which was extracted from the bladder by means of Sir Astley Cooper's forceps.

Presented by Mr. Aston Key, 1826.

1876   Oxalate of Calcium Calculus.

A small mulberry calculus somewhat quadrilateral in figure and measuring half an inch from side to side.

Presented by Mr. Aston Key.

1877   Oxalate of Calcium Calculus.

The half of a spherical calculus measuring two inches in diameter and having a dark brown tuberculated exterior. On section the central portion, which measures an inch across, is seen to be partially separated from the outer crust by a narrow interval which in the recent state was " occupied by soft animal matter."

Presented by Mr. B. Gibson.

1878   Oxalate of Calcium Calculus.

The two halves of a mulberry calculus, cubical in shape and measuring two inches diagonally. It weighs nearly four ounces and consists of a compact external layer enclosing a less dense material, which on section presents a radial arrangement of its structure.

From Joseph A., 1806.

1879   Urate of Ammonium Calculi.

Sections of three calculi, the largest of which measures half an inch in its longest diameter. They are formed of more or less compact concentric layers of a whitish material, which was found by Dr. Golding Bird to have a specific gravity of 1457 and to consist of urate of ammonium.   *Presented by Mr. Bransby Cooper.*

1880   Cystine Calculus.

The half of an oval calculus measuring an inch and a quarter in length and one inch in breadth. The cut surface is of light bluish-green colour and its exterior, which is slightly granular, is brownish. This calculus was analysed and described by Dr. Wollaston in the ' Transactions of the Royal Society ' in 1810 as consisting of cystic oxide, having been found by him amongst the collection of Guy's Hospital. It was again described and figured by Dr. Marcet in his work published in 1817. It was then of a brown colour. Its present blue tint dates from the year 1830 and perhaps earlier.

1881   Cystine Calculi.

A series of cystine calculi voided by a patient at intervals between the years 1814 and 1828. The last-passed calculus is elongated and shaped somewhat like an ear-drop, measuring an inch in length. The rest vary in size from a swan-shot to a pea, the three large upper ones being those first passed. They have a rounded shape, with broken crystalline surface, and are of various shades of blue and green. The long calculus passed in 1828 was described in the catalogue published in 1861 as of a greyish colour.

From Mr. B., who at the age of 23 was, in 1807, cut for stone. He subsequently suffered from pains in the loins, followed by hæmaturia and the passage of the calculi which form the preparation. The three stones earliest voided are described in Dr. Marcet's work.

Presented by Mr. Newington.

1882   Cystine Calculus.

A small oval calculus, barely half an inch in length, composed of cystine and having a specific gravity of 1777. The cut surface, which has been exposed to the light, is of an olive-green colour, whilst its granular exterior is light brown.

From a boy, æt. 12, of a "delicate and strumous habit," who in 1836 was suddenly seized with retention of urine. Mr. Aston Key removed the calculus from the urethra. When recent, it presented an "amber and translucent appearance," which had not altered when described by Dr. Golding Bird three years later. In 1861 there was a slight greenish hue on the cut surface. An analysis of the patient's urine was made by Dr. Golding Bird, and it was found to contain ·34 parts per thousand of cystine, which deposited in the form of hexagonal plates.

1883   Cystine Calculus.

The two halves of an oval calculus measuring two inches in length and an inch and a half in breadth, and weighing 840 grains. It is of a pale brown colour, its outer surface being finely granular and its cut section having a homogeneous waxy appearance. It has been analysed by Mr. Panting, who found it to be composed of cystine, with traces of phosphate of magnesium and calcium.

Charles P., æt. 38, was admitted under Mr. Howse for pain in the bladder and occasional difficulty in micturition. Thirteen years previously he passed a small calculus, and in the succeeding ten years he suffered from repeated attacks of renal colic, followed by the passage of small concretions. Three years before admission, an attack of colic occurred which was not followed by the passage of a stone, and during the last four months the patient had suffered from symptoms of vesical calculus. On admission the urine was found to be alkaline and to contain cystine crystals. Suprapubic lithotomy was performed, and the patient made a rapid recovery. *See Surgical Reps.* 1893, Case 240.

1884   Xanthic Oxide Calculus.

A small hard brown fragment of a calculus which weighed 339 grains and was removed from a child eight years of age by Professor Langenbeck at Hanover. It was analysed by Professor Stromeyer and found to consist of xanthic oxide. The original calculus was "a lustrous bright brown and composed of concentric separable layers, without any appearance of crystalline or fibrous texture; it was hard and had a wax-like lustre." *See Guy's Hosp. Rep.* 1842, p. 202.

1885   Phosphate of Calcium Calculus.

The half of a dense white pyriform calculus measuring an inch and a quarter in length and weighing 257 grains. Below is mounted a similar small button-like calculus, the convex polished surface of which was apparently lodged in a concave depression seen at the back of the larger stone. The concretions were analysed by Dr. Babington and found to consist of phosphate of calcium.   *Presented by Sir Astley Cooper.*

1886 Triple-phosphate Calculus.

A white friable calculus, oval in shape, measuring an inch and a half in length and weighing 205 grains. It presents a small central cavity in the walls of which are seen delicate crystals of triple phosphate, of which salt Dr. Babington found the calculus to be composed.

Presented by Mr. Aston Key.

1887 Fusible Calculi.

The halves of two calculi, the larger of which is elongated and measures an inch and a half in length. It presents a convex surface corresponding to a concavity in the smaller calculus. The concretions, which weighed rather more than half an ounce, were analysed by Dr. Babington and found to consist of fusible phosphates.

Presented by Sir Astley Cooper.

1888 Fusible Calculus.

A white oval calculus, measuring seven-eighths of an inch in length, and having a rough surface sprinkled with shining crystals. It was analysed by Dr. Golding Bird and found to consist of ammonio-magnesic phosphate with phosphate of calcium.

Presented by Mr. Bransby Cooper.

1889 Fusible Calculus.

The half of a small rounded calculus, weighing 40 grains, and composed of concentric layers of fusible phosphates. *Presented by Mr. Aston Key.*

1890 Fusible Calculus.

A calculus of irregular shape, measuring an inch and a quarter in length, and consisting of a body and nucleus of fusible phosphates surrounded by a thick friable incrustation of triple phosphate. It weighs 134 grains, and presents a groove at its smaller end corresponding to the neck of the bladder.

Presented by Mr. Aston Key.

1891   Fusible Calculus.

An oval calculus, measuring an inch in length, and composed of a white friable crust deposited at the ends and on one side of a more dense body and nucleus.

From a child, æt. 0, upon whom Mr. Aston Key performed lithotomy.

1892   Phosphatic Calculus.

An oval phosphatic calculus three quarters of an inch in length. It has a white rough surface.

James B., æt. 44, was admitted under Mr. Cock for stricture of the urethra and perinæal fistula. He died soon after admission, and at the autopsy the calculus which forms the preparation was found in a ragged cavity situated at the neck of the bladder behind the verumontanum. *See Insp.* 1858, No. 69.

1893   Phosphatic Calculi.

Several phosphatic calculi, the largest of which is oval in shape and measures half an inch in length.

From Thomas W., who suffered from urinary fistulæ opening in the perinæum and scrotum. Between thirty and forty calculi, similar to those forming the preparation, were removed from the sinuses by Mr. Cock in 1853.

1894   Phosphatic Calculus.

The half of a rounded calculus which weighed 94 grains and has a specific gravity of 1561. It consists of ammonio-magnesic phosphate, with phosphate of calcium. The central portion is of a purplish colour, while the outer layers are white and chalky.

From William P., æt. 3½, upon whom Mr. Bransby Cooper performed lithotomy.

1895   Phosphatic Calculi.

Two whitish phosphatic calculi of irregular shape and unequal size. The larger, which has a bent cylindrical figure, measures an inch and a half in length, and has a diameter of about a third of an inch. The smaller is flattened and roughly triangular.

From Charles K., æt. 7, upon whom Mr. Aston Key operated in 1820, removing the larger stone from the bladder and the smaller from a cavity in the perinæum communicating with the urethra.

1896   Phosphatic Concretion.

A flattened oval phosphatic concretion, a little less than half an inch in length, which was removed from the vaginal orifice of a vesico-vaginal fistula.

Presented by Mr. Bryant, 1864.

1897   Carbonate of Calcium Calculus.

The half of an elongated calculus, measuring an inch in width and nearly two inches in length. It consists of a white friable chalky material, found on analysis to be composed of carbonate of calcium with mere traces of alumina and carbonate of magnesia.

1898   Fragmentary Concretions of Urate of Sodium.

A numerous series of small fragments composed of urate of sodium.   *Presented by Dr. Wollaston,* 1825.

1899   Compound Mulberry Calculus.

The half of a hard dark brown calculus having an exceedingly irregular nodular surface. The polished section shews the central parts to be of a paler colour. On analysis the nucleus was found by Dr. Odling to consist of uric acid, while the outer layers were composed of oxalate of lime. *Presented by Mr. Cock,* 1851.

1900   Compound Mulberry Calculus.

The two halves of a calculus, weighing 365 grains, which was successfully removed by Mr. Aston Key. It was analysed by Dr. Babington and found to consist of a nucleus of uric acid surrounded by a thick coating of oxalate of lime. Its outer surface is nodular and covered with short thick spinous processes. *See Drawing*, No. 369 (91).

1901   Compound Calculus.

A rounded calculus, measuring an inch and a quarter in diameter, and having a white nodulated exterior. Its section exhibits a thick external layer of phosphate of calcium enclosing an irregular mass of oxalate of calcium with a central nucleus of uric acid.

1902   Compound Calculus.

A dark brown ovoid calculus, measuring two inches in length, and consisting of uric acid, surrounded by an irregular layer of white friable phosphates. It weighed 1610 grains.

Harry B., æt. 59, was admitted under Mr. Cock with symptoms of stone in the bladder, from which he had suffered for six years. On admission the urine was ammoniacal and contained pus. Mr. Cock performed lateral lithotomy with some difficulty on account of the size of the stone. Fifteen hours later the patient died, and at the autopsy the kidneys were found to be in a condition of suppurative nephritis. At the neck of the bladder was a pouch in which the calculus had been impacted. *See Insp.* 1856, No.119.

1903   Compound Calculus.

An irregularly spherical calculus, measuring two inches in diameter, and consisting of uric acid coated with a thick deposit of phosphates.

Presented by Mr. Callaway, Jun., 1856.

1904   Compound Calculus.

The half of a calculus, an inch and three quarters in diameter, composed of a white material, except for a small nucleus, which is of a darker colour and consists of uric acid. It has a finely granular surface, and its body is composed of phosphate of lime.

1905   Compound Calculus.

The half of an oval calculus, an inch and a half in length, and divided to shew a central portion consisting of uric acid and a smooth white outer coat of carbonate of lime. The central portion shews its outer layer to be partially detached and cracked, the interstices being filled with carbonate of lime.   *Presented by Mr. Bransby Cooper.*

1906   Compound Calculus.

The half of an oval calculus, an inch and a half in length, with a somewhat rough exterior, and seen on section to be composed of laminæ of a brownish substance enclosing a darker knuckle-shaped nucleus. It was analysed by Mr. Bransby Cooper, who found the nucleus to be oxalate of calcium and the body uric acid.

From George Q., æt. 68, upon whom Mr. Bransby Cooper performed lithotomy in 1850.

1907   Compound Calculus.

The half of a calculus of roughly quadrilateral figure, measuring two and a half inches in longest diameter and weighing $5\frac{1}{2}$ ounces. The polished surface shews it to be of dense structure and to consist of a darker body surrounded by a pale yellow crust, the former consisting of oxalate of calcium, the latter of uric acid. The exterior of the stone is coarsely nodulated and in parts exhibits a thin white deposit of phosphate of calcium. It was analysed by Mr. Bransby Cooper.

Presented by Mr. Bransby Cooper, 1832.

1908   Compound Calculus.

A light brown oval calculus, an inch and a quarter in length, and presenting on section a small nucleus of oxalate of calcium surrounded by uric acid.   Its exterior is finely granular.

1909   Compound Calculus.

A dense spheroidal calculus, measuring two inches in diameter, and having a small pale nucleus of oxalate of calcium surrounded by a brown radiate deposit of uric acid.   The exterior of the stone is partially encrusted with triple phosphates.   It weighs 518 grains and was analysed by Dr. Babington.

Presented by Mr. Aston Key.

1910   Compound Calculus.

The half of a rounded calculus, an inch in diameter, and weighing 142 grains.   It consists of a dark nucleus of oxalate of calcium, surrounded by a white crystalline deposit of ammonio-magnesic phosphate.   It was analysed by Mr. Bransby Cooper.

From William N., æt. 8 years, upon whom Mr. Bransby Cooper performed lithotomy in 1851.

1911   Compound Calculus.

The two halves of a rounded calculus, measuring an inch in diameter, and shewing on section its body to consist of a mulberry-shaped mass of oxalate of calcium crusted with white phosphates.   On the exterior the points of the stellate oxalate processes are seen to project slightly above the surface of the encrustation.

1912   Compound Calculus.

A roughly oval calculus measuring an inch and a half in its longest diameter, and consisting of a brown body of oxalate of calcium surrounded by a white friable crust of mixed phosphates. The calculus has a specific gravity of 1500. It was analysed by Dr. Golding Bird. *Presented by Mr. Bransby Cooper.*

1913   Compound Calculus.

A mulberry calculus of somewhat irregular shape, and measuring an inch and a quarter in its longest diameter. It consists of a whitish centre surrounded by a tuberculated outer crust of oxalate of calcium. On its exterior the intervals between the spicules of the oxalate salt are partially filled by a smooth white deposit of phosphate of calcium.

From William H., æt. 21, who was successfully lithotomised by Mr. Aston Key in 1848.

1914   Compound Calculus.

An oval calculus two inches in length, the body of which is composed of oxalate of calcium, and is surrounded by a soft white coating a quarter to half an inch in thickness, consisting of fusible phosphates.

From John H., æt. 8 years, who was successfully lithotomised in 1776.

1915   Compound Calculus.

A rounded calculus, 225 grains in weight, and measuring an inch and a quarter in diameter. It is composed of a whitish nucleus of urate of ammonium enclosed in an outer crust a quarter of an inch in thickness, which is brown in colour and consists of uric acid. The surface of the concretion is finely granular.

From William G., a boy, who was successfully cut for stone by Mr. Bransby Cooper in 1847.

1916   Compound Calculus.

An oval calculus measuring an inch and a half in length and having a brown granular surface. It was analysed

by Mr. Golding Bird, and found to have a specific gravity
of 1681, the nucleus being urate of ammonium, the
body uric acid, and the crust oxalate of calcium.

Presented by Mr. Bransby Cooper.

1917   Compound Calculus.

The half of a smooth oval calculus, measuring three
quarters of an inch in its longest diameter, and consisting
of a dark brown layer of oxalate of calcium enclosing a
lighter coloured nucleus of urate of ammonium.

From a child, æt. 8, upon whom Mr. Cooper Forster successfully performed lithotomy in 1851.

1918   Compound Calculus.

An oval calculus, an inch and a quarter in length, and
having a slightly granular pale brown exterior. Its
nucleus is composed of urate of ammonium, its body of
oxalate of calcium, and its crust of uric acid. The
calculus weighs 100 grains, and was analysed by
Mr. Bransby Cooper.

From a boy, æt. 11, who was operated upon for stone by
Mr. Bransby Cooper in 1848.

1919   Compound Calculus.

A large rounded calculus, an inch and three quarters in
diameter, the body of which is composed of oxalate of
calcium enclosing a nucleus of urate of ammonium, and
encrusted by a layer of fusible phosphates, in parts half
an inch in thickness. The calculus weighs 620 grains,
and was analysed by Mr. Bransby Cooper.

From William W., æt. 21, who was operated on for stone by
Mr. Bransby Cooper in 1841.

1920   Compound Calculus.

The half of a hard polished calculus, oval in shape and

measuring an inch and a quarter in length. The body of the concretion consists of oxalate of calcium, with a nucleus of urate of ammonium and an outer coating of fusible phosphates. The analysis was made by Mr. Bransby Cooper.

Removed after death from the bladder of a negro boy.

1921   Alternating Calculus.

The half of a mulberry calculus, with rounded outline, measuring an inch and a half in diameter. It has a nucleus of uric acid, surrounded by an irregular coat of oxalate of calcium. External to this is a deposit of uric acid with urate of ammonium, and the outermost crust is formed of oxalate of calcium. It was analysed by Dr. Owen Rees.   *Presented by Mr. Aston Key.*

1922   Alternating Calculus.

The half of a dark brown nodulated calculus, in the polished section of which is seen an oval nucleus of uric acid surrounded by a zone of oxalate of calcium with the characteristic "mulberry" projections. The intervals between these excrescences are filled by a deposit of uric acid which forms the crust of the stone.
Presented by Sir Astley Cooper.

1923   Alternating Calculus.

The half of an oval calculus an inch and three quarters in length, with a specific gravity of 1578, and weighing 587 grains. It was analysed by Dr. Babington and found to consist of oxalate of calcium with a nucleus and external coat of uric acid.
Presented by Mr. Bransby Cooper.

1924   Alternating Calculus.

A rounded calculus, measuring an inch and a half in diameter, and presenting on section a nucleus of uric acid surrounded by alternate layers of uric acid and oxalate of calcium. The outermost coat, which consists of oxalate of calcium, has a nodular surface somewhat resembling that of a mulberry.

1925   Alternating Calculi.

A series of fifteen calculi, for the most part spheroidal in shape, and somewhat altered by mutual contact. They vary but little in size, averaging an inch in diameter, and have a smooth white surface. One of them has been divided, and presents a porous brown centre surrounded by a paler and more compact crust. The analysis was made by Mr. Panting, who found the nucleus and outer zone to consist of pure uric acid and to be separated from each other by a layer of urate of ammonium.

John R., æt. 50, was admitted under Mr. Cock in a moribund condition and died on the following day. At the autopsy the kidneys, which were hydronephrotic, were in a condition of pyelonephritis. *See Insp.* 1859, No. 87.

1926   Alternating Calculus.

The half of an oval calculus which weighed 1848 grains, and measures two and a quarter inches in its longest diameter. It consists of an inner and outer layer of uric acid separated from each other by a thin layer of phosphatic material. The central part is of loose texture, while the outer zone is more compact.

Presented by Mr. Aston Key.

1927   Alternating Calculus.

A small oval calculus divided to shew a thin outer coating of triple phosphates covering a thicker layer of uric acid. Within this is a second layer of phosphates enclosing a small nucleus of uric acid.

1928 **Alternating Calculus.**

An elongated calculus, an inch and three quarters in
length and three quarters of an inch in thickness, consisting of a white friable material aggregated around
a firmer ovoid centre, which measures half an inch in
length, and is composed of alternate layers of uric acid
and phosphates. The outer coat, which has an irregular
rough exterior, consists of ammonio-magnesic phosphate
with phosphate of calcium.

1929 **Alternating Calculus.**

The half of an egg-shaped calculus, measuring two
inches in its longest diameter, and having a smooth
surface, except at its larger end, which corresponded
with the base of the bladder. It was analysed by
Mr. Brett and found to have a specific gravity of 1060,
with a nucleus consisting of uric acid and urate of
ammonium. The body consists of alternate layers of
a greyish and white material, the former having the
same composition as the nucleus, and the latter consisting of triple phosphate and phosphate of calcium.
The thin outer crust is a mixture of urate of sodium
and ammonium, with a minute proportion of uric acid.

Harry S., æt. 3, was admitted under Mr. Aston Key in
1826, with symptoms of stone in the bladder, from which he had
suffered for some months. About 12 weeks later he died, and at
the autopsy the bladder was contracted around the calculus, and
the kidneys were diseased. *See Insp.* vol. 7, p. 126.

1930 **Alternating Calculus.**

An oval calculus, weighing 702 grains, and measuring
an inch and three quarters in its longest diameter.
It presents on section a small dark brown nucleus
of oxalate of calcium surrounded by paler concentric
laminæ of uric acid. Its outer surface is finely granular,

R 2

and is sprinkled with minute shining crystals of oxalate of calcium.

From a lad, æt. 16, upon whom Mr. Cooper Forster performed lithotomy in 1859.

1931   Alternating Calculus.

An oval calculus, one inch and three quarters in length, and consisting of a white nucleus of oxalate of calcium, surrounded by thin alternate laminæ of uric acid and triple phosphate. The calculus weighs a little more than 7 drachms, and was analysed by Dr. Babington.

Removed by Mr. Aston Key in 1828 from the bladder of a young woman who had suffered from symptoms of vesical calculus for ten years.

1932   Alternating Calculus.

The half of a roughly quadrilateral calculus measuring an inch and a half in its longest diameter, and consisting of a nucleus of oxalate of calcium, surrounded by alternate layers of urate of ammonium and earthy phosphates.

1933   Alternating Calculus.

An oval calculus an inch and a half in length, having a smooth exterior, and consisting from within outwards of the following layers :—oxalate of calcium (nucleus) ; green cystic oxide ; urate of ammonium with fawn-coloured cystic oxide ; green cystic oxide ; urates, chiefly of ammonium ; alternate layers of urate of ammonium and oxalate of calcium. It was analysed by Dr. Golding Bird.

1934   Alternating Calculus.

An oval calculus one and a half inches in length and weighing 460 grains. Its nucleus and body consist of urate of ammonium surrounded by a crust a quarter of an inch in thickness, composed of alternate layers of oxalate and phosphate of calcium. Its outer surface is brown and granular. The analysis was made by Mr. Bransby Cooper.

From Joseph G., æt. 9, upon whom Mr. Bransby Cooper performed a successful lithotomy in 1845.

1935   Alternating Calculus.

A calculus of irregular shape, measuring an inch and a quarter in its longest diameter, and divided to shew its central portion to consist of alternate layers of urate of ammonium and phosphate of calcium outside which is a zone of oxalate of calcium between two layers of phosphate of calcium. The calculus, which weighs 280 grains, was analysed by Mr. Bransby Cooper.

From Mr. B., æt. 30, from whose bladder the stone was successfully removed in 1844 by Mr. Bransby Cooper.

1936   Alternating Calculus.

An oval calculus measuring an inch in length and having a nucleus of urate of ammonium with a white friable encrustation of fusible phosphates. The body of the concretion is composed of alternate layers of phosphate and oxalate of calcium. The calculus weighs 480 grains, and was analysed by Mr. Bransby Cooper.

From a child, æt. 8½ years, who was operated upon for stone by Mr. Philbrick in 1849.

1937   Mixed Calculus.

An oval calculus, measuring an inch and three quarters in its longest diameter, and thickly encrusted on one side with a whitish deposit consisting of triple phosphate externally, and a mixture of this salt with phosphate of calcium internally. The calculus itself has a nucleus of uric acid surrounded by oxalate of calcium, and external to this there is a thin dark band of phosphate of calcium. The concretion weighs 913 grains, and was analysed by Dr. Babington.

John W., æt. circa 40, was admitted under Sir Astley Cooper in 1804, with symptoms of vesical calculus. A fortnight later he died, and at the autopsy the left kidney, which was much enlarged, was found in a condition of suppurative nephritis, the right being atrophied. *See Old Museum Book,* p. 35 ; and *Prep.* 1760.

1938   Mixed Calculi.

A series of calculi, originally 142 in number, varying in size from a half to an eighth of an inch in diameter. Some are rounded, whilst others are facetted and roughly cubical or triangular. They have a smooth surface, and are of a brownish-white colour. On analysis by Dr. Golding Bird they were found to consist of uric acid with a small admixture of urate of sodium, urate of ammonium, and phosphate of calcium.

Removed by Sir Astley Cooper from Richard A., an adult, upon whom lithotomy was performed in 1811. Five years later he died shortly after a second operation, in which a stone of different character was removed.

1939   Mixed Calculus.

The halves of a calculus, bluntly triangular on section, and measuring two and a quarter inches along its base. It has a small eccentric nucleus of uric acid, and the same substance forms the dark brown granular crust. The body of the concretion consists of uric acid with urate of calcium. Its specific gravity is 1640. The analysis was made by Dr. Golding Bird.

Presented by Mr. Bransby Cooper.

1940   Mixed Calculus.

A small oval calculus with a whitish compact nucleus of urate of ammonium, surrounded by a brown crystalline

deposit, consisting of a mixture of uric acid with urate of calcium. It was analysed by Dr. Golding Bird.
Presented by Mr. Bransby Cooper.

1941   **Mixed Calculus.**

An oval calculus, measuring two inches and a quarter in its longest diameter, and consisting of a small nucleus of oxalate of calcium, surrounded by brownish concentric layers of uric acid with urate of ammonium. The concretion has a specific gravity of 1712, and its exterior is coarsely granular. It was analysed by Dr. Golding Bird.   *Presented by Mr. Bransby Cooper.*

1942   **Mixed Calculus.**

The half of a somewhat irregularly-shaped calculus, measuring an inch and three quarters in its longest diameter, and weighing 428 grains. It has a rough granular surface, and was found on analysis by Dr. Golding Bird to consist of urate of sodium mixed with a smaller quantity of urate of calcium.

From Mathew W., æt. 50, upon whom Mr. Bransby Cooper successfully performed lithotomy.

1943   **Mixed Calculus.**

The half of an oval calculus, measuring two inches in its longest diameter, and having a small nucleus of uric acid. The body of the stone is composed of the same substance, with the addition of urate of ammonium, and the thin brown crust consists of urate of ammonium and of calcium. The analysis was made by Dr. Golding Bird.   *Presented by Mr. Bransby Cooper.*

1944   **Mixed Calculus.**

A flattened oval calculus, measuring an inch in length and presenting a rough grey exterior studded with small shining crystals. The stone was analysed by Dr. Rees, who found that the thin outer coat consisted of oxalate of calcium with a minute proportion of phosphate of calcium, whilst the internal layer, which was less dense, consisted of uric acid with carbonate and phosphate of calcium.

George W., æt. 7, was admitted under Mr. Bransby Cooper in 1837, with symptoms of stone in the bladder, which had existed for six years. The calculus was removed by Mr. Bransby Cooper, the operation lasting 40 seconds. The patient was discharged from the hospital twenty days later perfectly well. *See Guy's Hosp. Reps.* 1837, p. 407.

1945   Mixed Calculus.

The half of a rounded calculus, which measured three inches in diameter, and weighed nearly 8 ounces. Its nucleus and the stellate portion around it consist of oxalate of calcium, the broad white soft zone external to this is phosphate of calcium, while from this band outwards the denser layers consist of the latter salt with variable proportions of ammonio-magnesic phosphate. The exterior of the stone is smooth, except for a partial granular deposit of the triple phosphate with a very small admixture of phosphate of calcium. The analysis was made by Dr. Babington.

Removed after death from the bladder of one of Mr. Bransby Cooper's patients.

1946   Mixed Calculus.

An oval calculus, three quarters of an inch in length, and having a brownish nucleus and body consisting of oxalate of calcium surrounded by a white encrustation of carbonate of calcium with triple phosphate. It was analysed by Dr. Owen Rees.

Presented by Mr. Bransby Cooper.

1947   Mixed Calculus.

A pear-shaped calculus, measuring three quarters of an inch in its longest diameter, and consisting of urate of ammonium with a thin whitish encrustation of triple phosphate and phosphate of calcium. It was analysed by Dr. Babington.

From James S., æt. 6, upon whom Mr. Bransby Cooper performed lithotomy in 1835.

1948 Mixed Calculus.

The half of a small nodular calculus of rounded outline, and weighing 22 grains. It is of a whitish colour and consists of a mixture of urate of ammonium with oxalate and carbonate of calcium.

From a boy, upon whom Mr. Bransby Cooper successfully performed lithotomy in 1835.

1949 Mixed Calculus.

A small oval calculus with a brown granular surface, divided to shew its nucleus consisting of uric acid, and its body consisting of the same substance with oxalate of calcium. Its external crust is formed of oxalate of calcium, beneath which is a thin whitish deposit of carbonate of calcium. The analysis was made by Dr. Golding Bird. *Presented by Mr. Bransby Cooper.*

1950 Mixed Calculus.

An oval calculus, measuring an inch and a quarter in its longest diameter, and having a specific gravity of 1545. On analysis by Dr. Golding Bird it was found that the body was composed of uric acid, whilst the thin outer whitish crust consisted of urate of calcium with phosphates. *Presented by Mr. Bransby Cooper.*

1951 Mixed Calculi.

The halves of two spherical calculi of almost equal size, measuring an inch and a quarter in diameter, and having a specific gravity of 1700 and 1680 respectively.

They were analysed by Dr. Golding Bird and Mr. Brett, and were found to consist chiefly of uric acid with traces of urate of ammonium, urate of calcium, and earthy phosphates.

From a male, æt. 57, upon whom Mr. Bransby Cooper performed lithotomy in 1835.

1952   **Mixed Calculus.**

An oval calculus, three quarters of an inch in length, with a body and nucleus consisting of urate of ammonium, surrounded by a whitish friable encrustation, an eighth of an inch in thickness, composed of uric acid and earthy phosphates.   It was analysed by Dr. Golding Bird.   *Presented by Mr. Bransby Cooper.*

1953   **Mixed Calculus.**

A rounded calculus, having a whitish smooth external surface, and measuring an inch and a quarter in diameter. It has a thick external layer of phosphate of calcium and phosphate of magnesium, and a nucleus of urate of ammonium.   The intermediate zone is composed of phosphate and oxalate of calcium.   The calculus weighs 209 grains, and was analysed by Mr. Bransby Cooper.

From A. W., æt. 52, who was successfully lithotomised by Mr. Bransby Cooper in 1841.

1954   **Mixed Calculus.**

The half of a large calculus, ovoid in shape, and measuring two and a quarter inches in its longest diameter.   The nucleus, which lies eccentrically, and the outer layers are of a compact brownish-yellow material, the parts between being of a less dense structure.   The outer surface is rough and scaly.   The stone was analysed by Mr. Brett and found to consist of a mixture of uric acid and urate of ammonium.

From Nathaniel H., upon whom the operation of lithotomy was successfully performed by Mr. Morgan in 1835.

1955   **Mixed Calculus.**

An oval calculus, measuring an inch and a quarter in its longest diameter, and consisting of a body composed of urate of calcium surrounding a nucleus of uric acid and urate of sodium. It was analysed by Dr. Golding Bird.   *Presented by Mr. Aston Key.*

1956   **Mixed Calculus. Internal Fracture.**

An oval calculus, rather more than half an inch in length, and having a brownish granular surface. It has been divided, and is seen on section to present a zigzag longitudinal fissure traversing its body and nucleus. The two fragments, which consist of urate of ammonium, are held together by an outer crust formed of uric acid with traces of earthy phosphates. The analysis was made by Mr. Brett.

From Benjamin B., æt. 3, a patient of Mr. Aston Key in 1835.

1957   **Mixed Calculus.**

The half of an oval calculus, measuring seven-eighths of an inch in length, and consisting of a grey body surrounded by a white crust. It was analysed by Dr. Golding Bird, and found to consist of phosphate of calcium with a small admixture of triple phosphate.

Presented by Mr. Aston Key.

1958   **Mixed Calculi.**

The halves of two calculi with fragments of a third which together weighed 1281 grains. They are moulded by mutual attrition and are composed of a white chalky substance, which was found by Dr. Babington to consist of fusible phosphates with some layers of pure phosphate of calcium.

Removed by Mr. Aston Key by lateral lithotomy, and thought by him to have been lodged in the prostate.

1959   Mixed Calculus.

The half of an oval calculus, measuring an inch and a half in length, and consisting of a compact white substance enclosing a small light brown nucleus. The latter consists of oxalate of calcium, whilst the body is composed of uric acid and phosphates. The analysis was made by Dr. Owen Rees.

From a girl, æt. 8, upon whom Mr. Aston Key successfully performed suprapubic lithotomy.

1960   Mixed Calculus.

The half of a quadrilateral calculus, measuring an inch and three quarters in length and having a dark brown granular exterior. On section it consists of compact concentric laminæ surrounding a less dense darker centre. It was analysed by Dr. Babington and found to consist of oxalate of calcium with some admixture of uric acid.   *Presented by Mr. Aston Key.*

1961   Mixed Calculus.

An oval calculus, measuring two and a half inches in length, and having a brown finely granular surface. Its thin external coat consists of uric acid with oxalate of calcium, the body and nucleus being composed of uric acid with urate of ammonium and triple phosphate. It was analysed by Dr. Owen Rees.

Presented by Mr. Aston Key, 1835.

1962   Mixed Calculus.

A rounded calculus, having upon its outer surface several nodular projections consisting of oxalate of calcium covered by a thin white deposit of triple phosphate. The cut section, which is compact and highly polished, shews a white nucleus, which was found on

analysis by Dr. Owen Rees to consist of urate of
ammonium. The oxalate of calcium which forms the
body of the calculus is mingled with uric acid.

From a boy, successfully lithotomised by Mr. Aston Key in
1834.

1963   Mixed Calculus.

The half of an irregularly oval calculus, measuring an
inch and three quarters in its longest diameter. It has
a central nucleus of uric acid surrounded by a browner
material consisting of a mixture of that substance with
triple phosphate, whilst its outer coat is composed of an
uneven deposit of triple phosphate. It was analysed by
Dr. Dowler.   *Presented by Mr. Aston Key.*

1964   Mixed Calculus.

The half of an oval calculus, measuring two and a half
inches in its longest diameter, and weighing 1390
grains. It was analysed by Dr. Owen Rees, who found
its thick white external crust to consist of triple phosphate, phosphate of calcium, and urate of ammonium.
The layers immediately beneath the crust were of uric
acid and urate of ammonium, whilst the body and
nucleus were of uric acid, with a less proportion of
urate of ammonium than in the intermediate layers.

Presented by Mr. Aston Key, 1834.

1965   Mixed Calculus.

The two halves of a sickle-shaped calculus, an inch and a
half long and half an inch in thickness. Its nucleus
consists of oxalate and urate of calcium in equal proportions, whilst the body and crust are formed of a mixture
of triple phosphate with urate of sodium and ammonium.
It was analysed by Dr. Golding Bird and Mr. Brett.

Presented by Mr. Aston Key, 1835.

1966   Mixed Calculus.

The half of an oval calculus, having a specific gravity of 1750, and measuring an inch and three quarters in length. It has a dark brown nodular external crust consisting of uric acid, urate of sodium, and traces of phosphate of calcium. Its cut section shews a nucleus of urate of ammonium and sodium surrounded by pale layers of urate of ammonium and phosphate of calcium which are separated from the external crust by a darker zone having the same composition as the nucleus. The analysis was made by Dr. Golding Bird.

Presented by Mr. Aston Key.

1967   Mixed Calculus.

A mulberry calculus, of pale colour and measuring an inch and a half in diameter. It was analysed by Dr. Golding Bird and found to consist of a mixture of oxalate of calcium and triple phosphate with traces of urate of sodium and ammonium.

Presented by Mr. Aston Key.

1968   Mixed Calculus.

The half of an ovoid calculus, measuring an inch and a half in length, and having a specific gravity of 1630. It presents a thin brown crystalline coating of oxalate of calcium, beneath which is a much thicker white layer consisting of a mixture of urate of ammonium, uric acid, ammonium chloride, phosphate of lime, carbonate of lime, and phosphate of ammonium and magnesium. Its nucleus is composed of uric acid, with urate of ammonium and traces of oxalate of calcium. The analysis was made by Dr. Golding Bird.

Presented by Mr. Aston Key, 1834.

1969   Mixed Calculus.

A calculus of irregular form, measuring three inches in its longest diameter and consisting from within outwards of the following layers : uric acid ; oxalate of calcium ; uric acid and oxalate of calcium ; ammonio-magnesic phosphate and urate of ammonium ; pure ammonio-magnesic phosphate in alternate layers, with a mixture of the same salt with urate of ammonium. The analysis was made by Dr. Pavy.

From Elias W., æt. 29, who was successfully lithotomised by Mr. Aston Key.

1970   Mixed Calculus.

A small oval calculus, three quarters of an inch long, and partially encrusted with a thin deposit of triple phosphate. The body of the concretion, which is of a brown colour, consists of uric acid and urates, whilst the paler nucleus is composed of urate of ammonium. The analysis was made by Mr. Bransby Cooper.

From Elizabeth S., a child, from whose bladder the calculus was removed by Mr. Hilton per urethram.

1971   Mixed Calculus.

The halves of an irregular calculus, measuring three quarters of an inch in longest diameter, and having a pale mammillated surface. It was analysed by Mr. Panting and found to consist of a mixture of oxalate of calcium and carbonate of calcium, with traces of phosphates and uric acid.

From a boy, æt. 14.

Presented by Mr. Hilton, 1863.

1972   Mixed Calculus.

The half of a dense white oval calculus, measuring two inches in length, and found on analysis by Dr. Odling to consist of a mixture of fusible phosphates with a small proportion of urate of ammonium.

Presented by Mr. Hilton, 1852.

1973   Mixed Calculus.

The half of a smooth ovoid calculus, measuring an inch and a quarter in its longest diameter. It consists of a brownish-yellow material aggregated around a small oval nucleus and presenting a concentric arrangement at its periphery. The calculus was analysed by Dr. Odling and found to consist of uric acid with urate of ammonium.   *Presented by Mr. Hilton, 1850.*

1974   Mixed Calculus.

A small oval calculus, consisting from without inwards of the following layers :—urate of ammonium with earthy phosphates ; oxalate of calcium ; urate of ammonium, sodium, and calcium ; urate of ammonium and uric acid. It was analysed by Mr. Bransby Cooper.

Presented by Mr. Hilton.

1975   Mixed Calculus.

A rounded calculus, measuring an inch in diameter, and shewing on section an intermediate zone of pure oxalate of lime combined externally with earthy phosphate and in the nucleus with uric acid and urate of ammonium. It was analysed by Dr. Odling.

Presented by Mr. Hilton, 1852.

1976   Mixed Calculus.

The half of an oval calculus, which measured a little more than an inch in its longest diameter, and weighed 105 grains. It has a dense brown nucleus of uric acid surrounded by a friable material consisting of a mixture of uric acid with oxalate of calcium. The outer crust is composed of urate of ammonium and phosphate of calcium and magnesium. The analysis was made by Mr. Panting.

From William W., æt. 7, upon whom Mr. Cock successfully operated for stone in 1854.

1977   Mixed Calculus.

The half of an oval calculus, an inch in length, and composed of a nucleus of oxalate of calcium, with a body consisting of a mixture of uric acid with urates of sodium and ammonium. There is a thin rough crust composed chiefly of earthy phosphates. The analysis was made by Dr. Odling. *Presented by Mr. Cock,* 1851.

1978   Mixed Calculus.

A spherical calculus, an inch in diameter, the central parts of which were found on analysis by Dr. Odling to consist of uric acid and urate of ammonium, whilst the rough white outer crust is formed of mixed earthy phosphates. *Presented by Mr. Cock.*

1979   Mixed Calculus.

A small calculus, the central portion of which consists of a mixture of uric acid and urate of ammonium, with traces of oxalate of calcium, whilst its outer layer is formed of pure oxalate of calcium. Its exterior is remarkably rough and covered with short spinous processes. It was analysed by Dr. Odling.
Presented by Mr. Cooper Forster.

1980   Mixed Calculus.

A small rounded calculus, measuring half an inch in diameter. It has a smooth surface and is of a light brown colour. It was analysed by Dr. Swayne Taylor and found to consist of uric acid with urate of ammonium.

Removed by Mr. Cooper Forster in 1858 from Thomas R., æt. 2½, who had suffered from symptoms of stone in the bladder for twelve months.

1981   Mixed Calculus.

The half of a dense pyriform calculus, measuring an inch and a half in length, and apparently formed around two nuclei. It has a brown spinous exterior, and on analysis by Dr. Odling was found to consist of a mixture of phosphate and oxalate of calcium with a small proportion of ammonio-magnesic phosphate.

Presented by Mr. Thomas Bryant, 1861.

1982   Mixed Calculus.

The half of an oval calculus, which weighed 984 grains and measured two inches in length and one inch and a half in thickness. Its section shews it to consist of an elongated body with a thick porous encrustation limited to one side. It was analysed by Mr. Panting and found to consist of a mixture of phosphates of calcium and magnesium with urate of ammonium.

From Ann C., æt. 35, from whose bladder the stone was removed by Mr. Bryant in 1864 through the dilated urethra.

1983   Mixed Calculus.

The half of an oval calculus, which measured four inches in its longest diameter and weighed 16 ounces. Its body, which is spherical and measures an inch and three quarters across, is formed of a light brown porous material. Its crust is compact and of a whitish colour. The stone was analysed by Mr. Panting who found the nucleus and body to consist of a mixture of oxalate of calcium and uric acid, and the crust to be composed of phosphate of calcium and magnesium with traces of carbonate of calcium.

Henry H., æt. 30, was admitted under Mr. Davies-Colley with a stone in the bladder, from symptoms of which he was said to have suffered "all his life." On admission a distinct tumour was felt above the pubes. Lithotomy was attempted and found impracticable on account of the great size of the stone, which was afterwards removed by suprapubic lithotomy. The patient was discharged in good health four months after the operation. *See Surgical Reps.* 1894, Case 181.

1984   Mixed Calculus.

A large ovoid calculus, measuring two inches and three quarters in length and apparently aggregated around two nuclei consisting of uric acid. The rest of the concretion is formed of the same substance with the addition of a small proportion of triple phosphate. Its outer surface is of a dark brown colour, and presents numerous small flat-topped elevations. It was analysed by Dr. Owen Rees, and was said to have been removed by Cheselden.   *Presented by Mr. W. H. Smith.*

1985   Mixed Calculus.

An ovoid calculus, measuring nearly two and a half inches in its longest diameter, and divided to shew a solid nucleus surrounded by less dense layers of a light brown material. The calculus exhibits a blunt tooth-like excrescence of a whiter substance, which also partially encrusts it and was found on analysis to consist of triple phosphate and phosphate of calcium. Its nucleus is composed of pure uric acid and its inter-mediate layers of this salt with traces of urate of ammonia and phosphate of calcium. The stone was analysed by Dr. Owen Rees.

1986   Mixed Calculus.

The half of an elongated calculus, measuring two and a quarter inches in length and three quarters of an inch in thickness. Its nucleus, which is placed eccentrically, consists of urate of calcium, and the rest of the con-cretion is formed of ammonio-magnesic phosphate and phosphate of calcium. This portion is white and friable, and before the blowpipe fuses into a colourless bead. The analysis was made by Mr. Golding Bird.

s 2

1987   Mixed Calculus.

A calculus of irregular shape, measuring an inch and a half in its longest diameter, and presenting on section a brown oval body with crenated outline, surrounded by a thick white deposit of phosphate of calcium. It was analysed by Mr. Brett, who found the nucleus to consist of oxalate of calcium, whilst the remainder of the body had a similar composition with traces of phosphate of calcium.

1988   Mixed Calculi.

Twelve smooth white calculi, varying in size from an inch and a half to half an inch in longest diameter They are of irregular shape and for the most part facetted. One of the calculi has been analysed and found to consist of fusible phosphates with traces of urate of ammonium.

Removed after death from a male patient.

1989   Mixed Calculus.

The half of a dense yellowish-brown calculus of oval shape, measuring two and a half inches in longest diameter. It has a specific gravity of 1400. It was analysed by Mr. Brett, who found the white portion of the external crust to consist chiefly of triple phosphate with a very small quantity of phosphate of calcium, so that it was only imperfectly fusible. The fawn-coloured portion of the crust consists of urate of ammonium with a small proportion of urate of calcium, uric acid, and earthy phosphates. The body of the calculus is composed of uric acid and urate of ammonium, with traces of ammonium and sodium chloride and earthy phosphates. The nucleus is nearly pure oxalate of calcium.   *Presented by Mr, Foaker.*

1990   Mixed Calculus.

The half of a dense oval calculus, measuring an inch and a half in its longest diameter, and presenting a brown finely-nodulated coating of oxalate of calcium. The rest of its structure is composed of alternate layers of uric acid and phosphate of calcium with triple phosphate.   *Presented by Mr. Blizzard.*

1991   Mixed Calculus.

The half of a small white calculus, oval in shape, and consisting of a nucleus of uric acid surrounded by a mixed coating of uric acid and oxalate of calcium. The surface is finely granular.

1992   Mixed Calculus.

The half of an irregularly shaped calculus, measuring nearly three inches in its longest diameter. It is partially coated by a thin whitish deposit consisting of uric acid with urate of calcium and ammonium. The nucleus consists of the same substances, whilst the body is composed of uric acid with urate of sodium and ammonium and phosphate of calcium. The calculus was analysed by Dr. Golding Bird and Mr. Brett.

1993   Mixed Calculus.

The half of an ovoid calculus, weighing 165 grains, and measuring an inch and a quarter in its longest diameter, Its surface is finely granular and its whitish nucleus consists of urate of sodium and phosphate of calcium with uric acid. The body of the stone, which is of a brownish colour, is composed of urate and phosphate of calcium.

Removed by lithotomy from a Hindoo child, æt. 3¾ years.

Presented by Mr. Fogerty, of Bombay, 1835.

1994   **Mixed Calculus.**

The half of a rounded calculus, measuring an inch in diameter, and consisting of a small nucleus of urate of ammonium surrounded by a thick crust of oxalate of calcium with traces of phosphate of calcium. Its outer surface is dark and nodular. The analysis was made by Mr. Bransby Cooper.

The calculus was removed from the bladder of a Malay girl by dilatation of the urethra.

1995   **Mixed Calculus.**

The half of a heavy compact calculus, having a finely polished section, and measuring two inches and three quarters in its longest diameter. It was analysed by Dr. Owen Rees, and found to consist of concentric layers of a mixture of uric acid with triple phosphate, and to have a thin outer shell of urate of ammonium. A small calculus is seen attached to the larger concretion.   *Presented by Mr. Camplin.*

1996   **" Dumb-bell " Calculus.**

Two calculi of unequal size, united by a short narrow neck, so as to somewhat resemble a "dumb-bell" or "bar-shot." The larger concretion, which is oval in shape and measures an inch and a quarter in its longest diameter, has been divided and exhibits a thick white external layer of fusible calculus, a nucleus of uric acid, and an intermediate zone composed of oxalate of calcium, with traces of uric acid and phosphates. "It was thought that the larger part had been lying in a pouch of the prostate, while the smaller portion remained in the bladder."

Henry N., æt. 20, was admitted under Mr. Callaway in 1854 with a calculus in the bladder. Perinæal lithotomy was performed, and the two calculi forming the preparation were separately removed. There was a third calculus weighing ten grains, which was extracted at the commencement of the operation. *See Guy's Hosp. Reps.* 1857, p. 351.

1997   Urinary Calculus.

About a third of a dense brown calculus, measuring an inch and a half in length, and having a rough granular surface. It weighed 360 grains.

Removed by Mr. Cock from the bladder of a woman in 1854, by dilatation and incision of the urethra.

1998   Urinary Calculus.

A flattened oval calculus, four inches long, three and a half inches broad, and an inch and a half thick, and weighing 16 ounces.

Removed by Sir Astley Cooper by lateral lithotomy. The calculus resisted all attempts to crush it. The patient died shortly after the operation.

1999   Urinary Calculi.

Nine smooth whitish calculi about the size and shape of pigeons' eggs.

Removed after death from the bladder of John G., æt 53, by Mr. Hingston in 1736.

2000   Cast of a Calculus.

A cast of a very large calculus, which weighed 25 ounces. It has a strongly marked oblique groove on what appears to have been its anterior part when in position.

From Sir Thomas Adams, æt. 81, who died in 1667. It was removed after death.

2001   Pretended Calculi.

Fragments of bone exhibited by an imposter as urinary calculi to obtain alms.

2002   Uric Acid Calculi passed by the Urethra.

A series of calculi, varying in size from a large pea to a swan-shot, passed at intervals during a period of seven months, the larger ones being those last passed. Beneath them is seen a collection of sandy material voided on two occasions in the previous year. The calculi and gravel are composed of uric acid.

Presented by Sir Astley Cooper, 1825.

2003   Calculus passed per Urethram.

An elongated oval calculus, measuring two inches and a quarter in length, and one inch and a quarter in thickness. It is brown in colour, spongy in structure, and has a very rough exterior.

It was voided through the meatus urinarius of a female.

Presented by Mr. Giraud.

2004   Calculi passed per Urethram.

A series of calculi voided on thirteen distinct occasions by the same person, and found, on analysis in 1835 by Dr. Golding Bird and Mr. Brett, to consist of urate of calcium, with traces of urate of ammonium and oxalate of calcium, the most external whitish crust being composed of phosphate of calcium.

2005   Calculi passed per Urethram.

Twenty-four calculi, rounded, facetted, and irregular in shape, and varying in size from one-half to a quarter of an inch in diameter. They were analysed by Mr. Brett and Dr. Golding Bird, and found to have a specific gravity of 1630, and to be composed of uric acid with a small proportion of urate of ammonium and of calcium.

Presented by Mr. Pearse in 1835, who stated that the concretions were " supposed originally to have formed one large calculus, and to have been split into fragments by the exhibition of Brandish's alkali." They were voided per urethram, and the patient quite recovered.

2006   Oxalate of Calcium Calculus passed per Ure-thram.

A small oval calculus about a quarter of an inch in diameter, whitish in colour, smooth on the surface, and composed of oxalate of calcium.

Voided by a female patient during micturition.

Presented by Dr. Rees.

2007   Calculus removed per Urethram.

The halves of a large ovoid calculus, weighing 740 grains, and measuring three inches in its longest diameter. It has a compact spherical nucleus, and appears to be composed of uric acid. The surface is coarsely granular.

Removed by Mr. Watson, of Stourport, through the meatus urinarius, from the bladder of a female.

Presented by Dr. Burne.

2008   Calculi removed per Urethram.

A series of uric acid calculi, ovoid in shape, and vary-ing in size from a large filbert to a pea. They were removed by Sir Astley Cooper from two male patients per urethram.

2009   Urinary Calculus removed from a Perinæal Abscess.

The halves of an oval calculus, measuring about half an inch in its longest diameter, which was removed from a perinæal abscess. The cut surface of the stone shews it to be of white colour and to possess a concentric structure.

2010   Calculus removed through a Recto-vesical Fistula.

Several fragments of a brownish-white calculus having a spiculated surface and a dark brown nucleus, removed during an operation for recto-vesical fistula.

2011   Calculus from the Scrotum.

A slender oval calculus, half an inch long, having a
brown colour and granular surface.

From Frank P., æt. 5, from whose scrotum the calculus was
removed by Mr. Poland in 1853. It was thought that the calculus
had made its way from the urethra by ulceration.

**2012   Phosphatic Calculus with Foreign Body as
Nucleus.**

An oval calculus, measuring two inches and a half in
length and an inch and a quarter in breadth. Its cut
section shews its long axis to be formed of a portion of
the stem of a long clay tobacco-pipe, around which is a
white loose deposit of phosphates.

Presented by Mr. Goodwin.

2013   Foreign Body from the Bladder.

A straw found in the bladder of a leper. It is four
inches and a half long, and is covered with a deposit of
phosphates, except the last inch, which projected into
the urethra.

For the history of the case see *Prep.* 1791.

Presented by Dr. Beaven Rake, 1886.

2014   Foreign Body from the Bladder.

A hair-pin, the greater part of which is embedded in a
mass of phosphatic deposit measuring two inches in
length and an inch in width.

Jeannette H., æt. 40, was admitted under Mr. Lucas for
pain and difficulty in micturition, with occasional hæmaturia.
"Ten months before ' she missed a hair-pin,' which she thought
might possibly account for her trouble." The urethra was dilated,
and the hair-pin was felt with its ends embedded in the mucous
membrane of the bladder. The calculus was withdrawn through
the urethra, and the patient was discharged well ten days after
the operation. *See Surgical Reps.* 1892, Case 192.

2015   Foreign Body from the Bladder.

A piece of a No. 3 gum elastic catheter, seven inches in length, and thinly coated with phosphatic deposit.

Henry II., æt. 52, was admitted under Mr. Poland in 1865. Five weeks before operation the instrument had slipped into the bladder. It was removed by lateral lithotomy.

2016   Foreign Body from the Bladder.

A withered stalk of parsley, the leaves of which are encrusted with phosphatic deposit.

Removed by lithotomy by Mr. Cock, 1852.

2017   Foreign Body from the Bladder.

A piece of leather boot-lace, together with fragments of phosphatic deposit with which it was encrusted.

From George II., upon whom Mr. Cock performed lithotomy in 1843.

2018   Foreign Body from the Bladder.

A piece of slate pencil two and a quarter inches in length, which formed the nucleus of a phosphatic calculus.

Thomas B., æt. 47, was admitted into Sidney Hospital, N.S.W., with symptoms of vesical calculus. Lateral lithotomy was performed by Mr. Alfred Roberts, and the patient made a good recovery. It was stated by the patient, and believed by some, that he had swallowed the pencil six months before its removal.

2019   Foreign Body from the Bladder.

A flattened distorted copper ring two and a half inches long and one inch and a half in width, which was removed by Sir Astley Cooper from the bladder of a woman. It is stated to have made its way into the bladder by ulceration through the wall of the vagina.

2020   **Bougie from the Bladder.**

A portion of a bougie with the phosphatic deposit with which it was encrusted.

John P., æt. 44, was admitted under Mr. Bransby Cooper in 1829, with symptoms of stone in the bladder. Lithotomy was performed, and some days later the patient died. *See Insp.* vol. 8, p. 141.

2021   **Catheter encrusted with Phosphates.**

A soft rubber catheter, the last four inches of which are thickly encrusted with white phosphatic deposit. On the reverse of the preparation are seen several masses of similar phosphatic deposit, which were broken off in extracting the catheter.

From Henry P., æt. 47, who was admitted, under Mr. Durham, with a perinæal fistula, in which a catheter had been retained four months. The fistula resulted from perinæal section performed for the relief of stricture of the urethra. *See Surgical Reports,* 1880, No. 99.

2022   **Catheter coated with Phosphates.**

A silver female catheter, the last two inches of which are thickly encrusted with a whitish phosphatic deposit, formed within fourteen days.

2023   **Concretions from a Case of Dermoid Tumour.**

A large number of irregularly shaped calculi consisting of soft white phosphatic material. In some of the concretions the deposit has been formed around hairs from a dermoid tumour ; and of many others, in which no hair can be seen, the filiform shape suggests a similar origin.

From a married lady, from whose bladder, on two separate occasions, portions of a dermoid tumour were removed by operation. *See Prep.* 1834.

Presented by Mr. Thomas Bryant.

SECTION XXXV.—DISEASES OF THE PROSTATE.

2024 Enlarged Prostate.

A prostate with the base of the bladder mounted to shew a globular tumour obstructing the neck of the bladder, and consisting of an adenomatous enlargement of the middle lobe of the prostate. The lateral lobes of the gland are also enlarged, and the organ measures nearly two inches in transverse diameter. The walls of the bladder are hypertrophied and its mucous membrane is partially destroyed by ulceration. Histological examination shews the middle lobe to consist of normal prostatic tissue with dilated glands.

William W., æt. 65, was admitted under Mr. Aston Key in 1839 for retention of urine with pain in the bladder and in the region of the kidneys. The patient died three weeks after admission, and at the autopsy "minute suppurating points of a whitish colour were noticed distributed throughout the greater part of the structure of the kidneys." *See Insp.* vol. 29, p. 49.

2025 Enlarged Prostate.

A bladder, laid open to shew its wall hypertrophied so as to measure half an inch in thickness and its neck obstructed by a globular enlargement of the middle lobe of the prostate of about the size of a cherry.

The lateral lobes of the gland are also hypertrophied, the right being a little larger than the left. On the reverse of the specimen the ureters are seen to be considerably dilated.

2026   Enlarged Prostate, perforated by a Catheter.

A prostate with the base of the bladder mounted to shew its middle somewhat enlarged and perforated to the right of the middle line. The lateral lobes are also enlarged.

Henry P., æt. 51, was admitted under Dr. Bright in 1836 for phthisis, having previously suffered from difficulty in micturition for which he had often been catheterised. He died three weeks after admission, and at the autopsy the interior of the bladder was found to be ulcerated and partially coated with false membrane. The kidneys were in a condition of pyelo-nephritis. *See Insp.* vol. 21, p. 97.

2027   Enlarged Prostate, tunnelled by a Catheter.

A large and sacculated bladder with the prostate. The gland is greatly hypertrophied and measures two and a half inches in transverse diameter. Its middle lobe, which projects prominently into the neck of the bladder, is pyriform in shape and rather more than an inch from side to side. A red rod indicates a perforation passing through the base of the middle lobe of the prostate and entering the bladder just beyond it.

2028   Enlarged Prostate.   Multiple Perforations.

A prostate, the lobes of which are greatly enlarged so as to measure two and a half inches in transverse diameter. The middle lobe forms a triangular prominence obstructing the neck of the bladder and presents several perforations produced by instruments.

2029   Enlarged Prostate, perforated by a Catheter.

An enlarged prostate, three inches in transverse diameter and having a middle lobe which measures an inch and a half from side to side and projects three quarters of an inch above the base of the bladder. The middle lobe is separated into two halves by a median furrow, and the left half presents a perforation produced by a catheter.

William R., æt. 77, was admitted under Mr. Cock in 1844 for retention of urine, which was relieved by the passage of a flexible catheter. For four years before admission he had suffered from difficulty in micturition, requiring the occasional use of an instrument. After his discharge from the hospital his urinary trouble gradually increased until his death at the age of 89.

Presented by Mr. Smith, of Crawley, 1856.

2030   Enlarged Prostate. Cystitis.

A sagittal section through the bladder, prostate, and penis, shewing the middle lobe of the prostate considerably enlarged, perforated by a catheter and excavated by ulceration. The mucous membrane of the bladder presents a polypoid surface as the result of extensive inflammation and ulceration. Histological examination shews that the appearances are due to chronic and recent inflammation, with no evidence of tubercle or malignant growth.

2031   Enlarged Prostate treated by Castration.

The base of a bladder with the prostate, which is enlarged so as to measure two and a half inches in transverse diameter. The middle lobe projects half an inch above the base of the bladder, the wall of which is considerably hypertrophied. The transverse section through the prostate shews a cystic dilatation of some of the ducts.

John W., æt. 73, was admitted under Mr. Davies-Colley for difficulty in micturition attributed to enlargement of the prostate. Double castration was performed, and the patient was discharged relieved. A few weeks later his symptoms returned, and, increasing in severity, proved fatal eight months after the operation.

At the autopsy the bladder was found to be inflamed and the kidneys were in a condition of pyelo-nephritis. *See Insp.* 1895, No. 501.

2032   Abscess of the Prostate.

A prostate with the adjacent portion of the bladder mounted to shew the former converted into a cavity with ragged walls, which in the recent state contained pus. The mucous membrane of the bladder has been destroyed by ulceration, exposing the muscular and, in parts, the serous coat of the viscus. The right ureter is much dilated.

From John P., æt. 53, who at intervals during the four years preceding his death passed blood and phosphatic deposit with his urine. At the autopsy the right kidney was found to be filled with soft cretaceous material. *See Prep.* 1736; and *Note Book*, i. p. 154.

2033   Tuberculosis of the Prostate.

A bladder, with the prostate and urethra laid open to shew the prostate somewhat enlarged and infiltrated with caseous deposit, which in parts has softened so as to form small cavities. The mucous membrane of the urethra is eroded by ulceration, and beneath it are seen miliary tubercles. On the reverse of the specimen the vesiculæ seminales are seen to be enlarged by caseous deposit.

Cornelius D., æt. 40, was admitted under Mr. Birkett with a perinæal abscess, which was opened and caused a urinary fistula. Five months later he died, and at the autopsy tuberculosis was found of the vertebræ, lungs, testes, kidneys, and intestines. *See Insp.* 1873, No. 137.

2034   Tuberculosis of the Prostate.

A transverse section through a prostate gland which is enlarged so as to measure an inch and three quarters in diameter. The gland and the vesiculæ seminales are infiltrated by a uniform caseous deposit, which in the centre of the prostate has softened so as to produce a small irregular cavity. Histological examination shews the organ to be affected by caseating tubercle.

John H., æt. 44, was admitted under Mr. Jacobson with tuberculous disease of the right epididymis associated with phthisis. The testicle was removed, and the patent died about four weeks later from his pulmonary disease. At the autopsy there was tuberculous ulceration of the large and small intestine and the right vas deferens was caseous. *See Insp.* 1802, No. 3.

2035   Fibro-myoma of the Prostate.

A hypertrophied bladder into the neck of which projects a trilobed tumour produced by adenomatous enlargement of the middle lobe of the prostate. The lateral lobes are little if at all enlarged. Histological examination shews the enlargement to be due to increase of the fibro-myomatous tissue of the organ.

Thomas G., æt. 60, was admitted under Mr. Hilton for urinary symptoms due to enlargement of the prostate, and died one month later from uræmic coma. At the autopsy the bladder was found to be hypertrophied and the pelves of the kidneys were dilated. The glandular structure of the kidneys was atrophied, their surfaces granular and their capsules firmly adherent. *See Insp.* 1863, No. 95.

2036   Adenoma of the Prostate.

A bladder with the proximal portion of the penis laid open to shew the prostatic urethra dilated and occupied by a globular mass measuring an inch in diameter. The mass has a broad attachment to the posterior wall of the urethra, and histologically is seen to have the structure of an adenoma of the prostate. At the neck of the bladder to the right of the middle line is seen the opening left after puncture per rectum. The bladder is dilated and its walls are much hypertrophied.

William B., æt. 64, was admitted under Mr. Birkett for fracture of the leg. Retention of urine supervened and the bladder was punctured from the rectum. Six days later the patient

274 DISEASES OF THE PROSTATE.

died, comatose, and at the autopsy the kidneys were found to be in a condition of advanced interstitial nephritis. *See Insp.* 1873, No. 61.

2037 Cystic Adenoma of the Prostate.

A bladder laid open to shew a hemispherical tumour about an inch in diameter arising from the left lobe of the prostate. It has a convoluted structure and its surface is rough and denuded of mucous membrane. The bladder, which is occupied by four facetted calculi, is considerably hypertrophied, and at its summit there is a cavity about as large as a cherry which in the recent state contained pus. On the reverse of the specimen the ureters are seen to be much thickened and dilated. Histological examination of the tumour shews it to have a fibro-myomatous structure with cystic dilatation of the glands.

Edward B., æt. 65, was admitted under Mr. Howse for incontinence of urine, having suffered from difficulty in micturition for three years. The prostate was found to be enlarged and tender, and on bimanual examination under chloroform calculi were detected in the bladder. The patient died a few days after admission from pyelo-nephritis. *See Insp.* 1887, No. 328.

2038 Carcinoma of the Prostate.

A dilated and hypertrophied bladder laid open to shew the prostate considerably enlarged and uniformly infiltrated by a growth having the histological characters of a spheroidal-celled carcinoma with scanty stroma. Over the growth the mucous membrane of the bladder presents several small nodular excrescences.

Robert H., æt. 67, was admitted under Mr. Poland for urinary symptoms from which, two days later, he died. At the autopsy no secondary deposits of growth were found. *See Insp.* 1862, No. 108.

2039 Carcinoma of the Prostate.

A prostate with the base of the bladder mounted to shew the gland considerably enlarged by a deposit which histologically has the characters of a spheroidal-celled carcinoma with scanty stroma and partial colloid change. On the reverse of the specimen the growth is seen, by extension, to have produced a mass as large as a chestnut surrounding and infiltrating the membranous portion of the urethra.

John B., æt. 68, was admitted under Mr. Cock with difficulty in micturition for which he had been in the habit of using a catheter. He died from pyæmia, and at the autopsy secondary deposits of carcinoma were found in the liver. There were abscesses in the right kidney. See Insp. 1865, No. 100.

2040   Carcinoma of the Prostate.

The base of a bladder with the prostate mounted to shew the gland considerably enlarged in all its dimensions. The middle lobe is prominent, and the mucous membrane covering the trigone of the bladder is tufted with villous processes. The wall of the bladder is much hypertrophied, and on the reverse of the specimen the vesiculæ seminales are seen to be embedded in the growth and the left ureter to be considerably dilated. Histologically the tumour has the characters of a spheroidal-celled carcinoma with scanty stroma.

Thomas B., æt. 78, was admitted under Mr. Jacobson for irritability of the bladder due to enlargement of the prostate. A catheter was passed and six ounces of clear residual urine were withdrawn. Hæmaturia and cystitis supervened, and the patient died with symptoms of uræmia four weeks after admission. At the autopsy both kidneys were found to be hydronephrotic, the right presenting points of suppuration beneath the capsule. See Insp. 1894, No. 23.

2041   Carcinoma of the Prostate and Kidney. Hydronephrosis.

A bladder with the right kidney and ureter. The bladder has been divided to shew in the situation of the prostate a large mass of soft white growth ulcerating into the base of the bladder and completely occluding the lower end of the ureter. The ureter at

T 2

this part is enormously dilated so as to measure five inches in circumference. The kidney is converted into a hydronephrotic sac seven inches in length, and attached to its thin fibrous walls are numerous deposits of malignant growth, some of a sessile and some of a polypoid character. Histological examination shews the growth to be a spheroidal-celled carcinoma.

Walter W., æt. 47, was admitted under Mr. Howse for difficulty in micturition and hæmaturia, due to an enlarged prostate. A tumour was felt in the region of the right kidney. A catheter was tied into the bladder, and the patient died four days after admission. At the autopsy secondary deposits of growth were found in the lumbar lymphatic glands and in the lungs and liver. See Insp. 1898, No. 75.

2042   Carcinoma of the Prostate.

The two halves of a bladder with the prostate divided by sagittal section. The gland is enlarged so as to measure four inches in its longest diameter, the enlargement being due to uniform infiltration by a soft white growth having the histological characters of a spheroidal-celled carcinoma. There is an irregular excavation in the centre of the growth communicating freely with the lower part of the rectum. The ureters are dilated, and the left is seen to be formed of two branches which unite about two inches above its termination in the bladder.

Richard M., æt. 74, was admitted under Mr. Howse with several inches of bowel prolapsed through an artificial anus produced by inguinal colotomy performed four months previously for supposed cancer of the rectum. The patient stated that he some time before had undergone a suprapubic operation for the removal of a vesical tumour. The prolapsed bowel was removed, and the patient died three weeks after the operation. No secondary deposits were found at the autopsy. See Insp. 1896, No. 355.

2043   Sarcoma of the Prostate.

A bladder, apparently from a young subject, with the prostate, which is greatly enlarged so as to measure an

inch and a half in transverse diameter. The enlargement is due to a uniform infiltration of the gland by
growth which also involves the neck and trigone of the
bladder, producing a warty condition of the overlying
mucous membrane. In the prostatic urethra the growth
forms nodular excrescences obstructing the channel.
Histologically the tumour is a sarcoma consisting of
short spindle and round cells.

2044   **Sarcoma of the Prostate.**

A sagittal section through the prostate and bladder of a
child, shewing the gland to be enlarged so as to measure
an inch and a half in length. Histological examination
shews the enlargement to be due to a uniform infiltration of the organ with a sarcomatous deposit the
cells of which are round and short spindle in shape.
There are areas of mucoid degeneration.

2045   **Sacculation of the Prostate.**

A bladder with the urethra in the penile portion of
which is a stricture. On either side of the verumontanum a bristle has been passed along one of the
prostatic ducts, and is seen, on the reverse of the specimen, to enter a cystic cavity situated in the lateral lobe
of the gland. These cysts are of equal size and measure
about three quarters of an inch in longest diameter.

Joseph V., æt. 53, was admitted under Dr. Back in 1834
with pericarditis and pleurisy, from which three weeks later he
died. "It was known that his urethra had been imperfect at
least for many years." At the autopsy the kidneys were found to
be granular.

2046   **Sacculation of the Prostate.**

A bladder considerably hypertrophied and affected by
membranous inflammation, laid open to shew in the
prostatic urethra just in front and to the right of the

verumontanum a circular orifice through which a blue rod has been passed into a sacculated cavity occupying the interior of the prostate.

Elijah H., æt. 47, was admitted under Mr. Davies-Colley for extravasation of urine. Cock's operation was performed, and shortly afterwards the patient died. At the autopsy a stricture was found in the spongy portion of the urethra. *See Insp.* 1885, No. 26.

2047   Sacculation of the Prostate.

A somewhat hypertrophied bladder with the prostate and proximal portion of the urethra, opened by frontal section to shew in the prostatic urethra just in front and to the right of the verumontanum a round opening a line in diameter communicating with a cavity in the prostate. On the lateral aspect of the preparation the wall of the cavity has been partially removed to shew that it measures about an inch and a half in diameter, is loculated, and has a smooth fibrous wall. In the bulbous portion of the urethra there is a stricture.

George G., æt. 38, was admitted under Dr. Shaw with lobar pneumonia and delirium tremens, from which two days later he died. At the autopsy the pelves of the kidneys were found to be dilated. *See Insp.* 1893, No. 160.

2048   Sacculation of the Prostate.

A bladder with a portion of the urethra laid open to shew in each lateral lobe of the prostate a smooth-walled cyst measuring rather more than an inch in diameter. There is a stricture at the commencement of the spongy portion of the urethra, and at this point a blue rod has been inserted into a false passage which is seen to terminate in the prostatic urethra between the two sacculi. The wall of the bladder is much hypertrophied.

From a patient who for several years suffered from considerable difficulty in micturition. He was accustomed to obtain relief by pressing upon his perinæum, by which act it was thought that he emptied the prostatic sacculi, which being distended by urine obstructed the urethra.

Presented by Mr. Griffiths.

2049   Sacculation of the Prostate.

The base of a bladder seen from behind and mounted to shew the prostate converted into a congeries of thin-walled intercommunicating cysts, the largest of which measures an inch and a half in transverse diameter. In one of the loculi there is a small brown calculus and on the reverse of the specimen a similar calculus is seen at the orifice of one of the dilated ducts opening into the urethra. The bladder is hypertrophied and sacculated, and there is a false passage tunnelling the prostate.

From a patient, the subject of stricture of the urethra.

2050   Dilatation of the Prostatic Urethra.

A prostate with the adjacent parts of the bladder and urethra. The urethra presents a stricture and a false passage at the junction of its membranous and bulbous portions. Between the stricture and the neck of the bladder the urethra is greatly dilated, forming a pouch the walls of which are composed of a thin shell of atrophied prostatic tissue. The verumontanum is seen in a recess behind a fold of mucous membrane at the neck of the bladder.

From John C., æt. 34, who was under the care of Dr. Golding-Bird in 1849 for cardiac disease.

2051   Dilatation of the Prostatic Urethra.

A prostate with the base of the bladder mounted to shew the prostatic urethra dilated so as to form a pear-shaped cavity large enough to admit the terminal phalanx of the little finger. The communication between this cavity and the bladder is marked by a thin transverse fold of mucous membrane in the situation of the middle lobe of the prostate. The ureters are normal.

From a patient who suffered from stricture of the urethra.

2052   **Prostatic Calculi.**

A series of six calculi, varying from an eighth of an inch to half an inch in diameter. They have a compact white structure and are facetted.

Removed from the prostate after death.

2053   **Prostatic Calculi.**

Four small white calculi, each measuring about a quarter of an inch in diameter. They are polygonal in shape and moulded by mutual contact.

Presented by Mr. Pugh.

2054   **Prostatic Calculi.**

A series of white concretions varying in size from a grain of sand to a quarter of an inch in diameter. They are of irregular shape, hard, and fragmentary in appearance. One was analysed by Mr. Panting, and found to consist of calcium, magnesium, and ammonium in combination with phosphoric, carbonic, and oxalic acids probably as triple phosphate, with oxalate and carbonate of calcium. There is no uric acid, but there is a small amount of organic matter.

Removed from the prostate gland on two separate occasions by Mr. Aston Key in 1828.

2055   **Calculus in the Prostate.**

A contracted and hypertrophied bladder laid open to

shew upon the posterior wall of the prostatic urethra a small opening through which is seen the rough surface of a brown calculus firmly lodged in a sacculus of the prostate.

William R., æt. 52, was admitted under Mr. Birkett for stricture of the urethra and perinæal fistulæ, from which he had suffered for nine years. He died four months after admission from ascending nephritis. *See Insp.* 1854, No. 139.

2056   Calculus in the Prostate.

A prostate containing in each lateral lobe a smooth-walled sacculus, in one of which is lodged a white calculus of irregular shape, measuring half an inch in its longest diameter.

2057   Calculi in the Prostate.

A bladder, with the urethra opened from behind to shew a rough cylindrical calculus, about half an inch in length, impacted at the neck of the bladder. The prostate has been incised in several places and exhibits similar calculi lodged in dilatations of its ducts.

Presented by Mr. Aston Key.

2058   Concretions in the Prostate.

Thin sections of a prostate gland, dried, and immersed in turpentine to shew numerous minute concretions of a dark brown colour contained within the substance of the organ.

2059   Concretions in the Prostate.

A portion of a prostate gland laid open to shew a considerable number of small brown concretions contained within its ducts.

Section XXXVI. — DISEASES OF THE VASA DEFERENTIA AND VESICULÆ SEMINALES.

2060   Absence of Vasa Deferentia.

A bladder, seen from behind, mounted to shew atrophy of the prostate and vesiculæ seminales and absence of the vasa deferentia. There is recent lymph in the recto-vesical pouch and the interior of the bladder is ulcerated.

Thomas M., æt. 18, was admitted under Mr. Aston Key in 1833 with urinary symptoms from which he had long suffered. Ten months later he died, and at the autopsy the testes were found to be small, the left kidney was rudimentary and its ureter deficient at the lower end. Death resulted from suppurative peritonitis.

2061   Calcification of the Vas Deferens.

Portions of a pair of vasa deferentia, dried and preserved to shew partial calcification of their coats.

John W., æt. 89, was admitted under Mr. Morgan in 1844 for traumatic cellulitis of the left leg. Ten weeks later he died, and at the autopsy the aorta was found to be very atheromatous, and there was a hæmatocele on either side. *See Insp.* vol. 33, p. 2 ; and *Preps.* 1217 (50) and 1384 (46) [2nd Edit.].

2062 Myxo-Lipoma of the Spermatic Cord.

A mass of fibro-fatty tissue at the lower end of which is seen a testicle with its vas deferens. Attached to the mass at the upper end of the spermatic cord is a partially encapsuled deposit of globular figure measuring four inches in diameter. Histological examination of the encapsuled tumour shews it to have the structure of a myxo-lipoma.

From a man, æt. 35, who had noticed a swelling in the scrotum gradually increasing in size for five years. It was partially excised, and after death the fibro-fatty tissue in which the growth lay was found to extend along the inguinal canal into the abdomen, its upper end reaching almost to the right kidney. *See Trans. Path. Soc.* vol. 40, p. 282.

Presented by Mr. Bryant, 1889.

2063 Carcinomatous Deposit in the Spermatic Cord.

A portion of a spermatic cord which is infiltrated by a mass of soft growth having the histological characters of a spheroidal-celled carcinoma.

2064 Myxo-Sarcoma of the Spermatic Cord.

The half of a dense globular tumour measuring three and a half inches in diameter, which was dissected out from the spermatic cord. Its cut surface shews it to consist of white interlacing fibres. Histologically it is a sarcoma nearly all the cells of which have undergone myxomatous change.

George W., æt. 26, was admitted under Mr. Hilton in 1860 for a tumour in the groin which had existed as long as the patient could remember and had gradually increased in size. The tumour was tapped and an ounce of clear fluid withdrawn. Subsequently the growth was excised. It was found to be attached to the spermatic cord, the testis being healthy. *See Drawing,* 420 (5); and *Wax Model,* 100 (10).

2065   Melanotic Sarcoma of the Spermatic Cord.

A testis with the spermatic cord, dissected to shew close to the globus major an ovoid nodule of black growth measuring three quarters of an inch in its longest diameter. Histologically the deposit is a melanotic sarcoma with round and oval cells.

James II., æt. 60, was admitted under Mr. Birkett with numerous small melanotic tumours scattered over his body. Eighteen months previously he had noticed a swelling of his left foot, shortly followed by enlargement of the thigh and of the inguinal glands. The patient died four months after admission, and at the autopsy melanotic growth was found in the heart and peritoneum and in the abdominal and thoracic lymphatic glands, *See Insp.* 1854, No. 56; *Prep.* 1400 (25) [2nd Edit.]; *Drawings,* 188 (26), 188 (27), and 463 (5); and *Wax Models,* 293 (5) and 293 (6).

2066   Melanotic Sarcoma of the Spermatic Cord.

A left testicle with a portion of the spermatic cord mounted to shew, about an inch above the globus major, a considerable mass of melanotic growth measuring two and a half inches in its longest diameter. The upper part of the tumour is soft and diffluent, while the lower portion is more solid and presents one or two small white nodules in its substance. Histologically the growth is a spindle-celled sarcoma and contains much black pigment.

Charles D., æt. 70, was admitted under Mr. Lane in 1891 for a painless tumour in the left groin which had been noticed for six months. He had worn a truss for nine years. Three years before admission the patient's left eye was excised at the London Hospital. The parts shewn in the preparation were removed together with an enlarged iliac gland, and the patient was discharged well shortly after the operation. *See Surg. Reps.* vol. 156, Case 51.

2067   Tuberculosis of the Vesiculæ Seminales.

The upper end of a vas deferens with the vesicula seminalis, the latter having been divided to shew in its tubes a deposit of caseous tubercle.

Blaxson B., æt. 48, was admitted under Dr. Babington with delirium and partial coma, and died a month later from tuberculous meningitis. At the autopsy miliary tubercles were found in the lungs and kidneys, the intestine was ulcerated, and the mesenteric and mediastinal glands were caseous. *See Insp.* 1854, No. 51.

2068   Tuberculosis of a Vesicula Seminalis.

The base of a bladder, seen from behind, mounted to shew the right vesicula seminalis infiltrated with a yellow softening deposit and enlarged so as to be about four times the size of its fellow. Histologically the deposit is seen to consist of caseating tubercle with giant cells.

William N., æt. 58, was admitted under Dr. Rees with symptoms of nephritis, and died eight months later. At the autopsy the kidneys, testes, and prostate were found to be affected with caseating tubercle, and there was miliary tuberculosis of the lungs and liver. *See Insp.* 1857, No. 105.

2069   Tuberculosis of a Vesicula Seminalis.

The base of a bladder, seen from behind, shewing the right vesicula seminalis greatly enlarged by a deposit of caseous tubercle. The left vesicula seminalis appears to be normal.

George G., æt. 36, was admitted under Dr. Gull for phthisis, from which, two months later, he died. At the autopsy miliary tubercles were found in the liver and kidneys. The right testis and epididymis were caseous. *See Insp.* 1864, No. 64; and *Preps.* 2099 and 2144.

2070   Tuberculosis of the Vesiculæ Seminales.

The base of a bladder with the prostate, seen from behind. A section has been made through the prostate and vesiculæ seminales to shew them to be infiltrated with caseous tubercle, the disease being most advanced in the right vesicula. The left ureter is dilated, and the right is thickened from tuberculous deposit beneath its mucous membrane. On the reverse of the specimen the interior of the bladder is seen to be eroded by ulceration.

James C., æt. 45, was admitted under Mr. Lane for hæmaturia, from which he had suffered at intervals for two years. Eleven days later he died, and at the autopsy tuberculous deposit was found in the lungs, intestines, and right kidney. The left kidney was in a condition of pyelo-nephritis. *See Insp.* 1890, No. 370.

2071   Calculi from the Vesiculæ Seminales.

Five white spherical calculi the size of millet-seeds which were found in the vesiculæ seminales.

Richard F., æt. 46, was admitted under Dr. Addison for suppuration in the chest-wall, and died from multiple abscesses in the brain. At the autopsy a stricture was found in the urethra; the vesiculæ seminales appeared to be normal. *See Insp.* 1854, No. 222.

Section XXXVII.—DISEASES OF THE TESTIS AND EPIDIDYMIS.

Hydatids of Morgagni : 2072.
Undescended Testis : 2073–2077.
Strangulation : 2077–2079.
Abscess : 2080–2082.
Calcification : 2083.
Osseous Deposit : 2084.
Hernia Testis : 2085, 2086.
Syphilis : 2087–2095.
Tubercle : 2096–2103.
Epithelioma : 2104.
Carcinoma : 2105–2111.
Sarcoma : 2112–2131.
Mixed Tumours : 2132–2140.
Cysts : 2141–2145.

2072   Hydatids of Morgagni.

A testis mounted to shew four small pedunculated cysts attached to the globus major of the epididymis.

2073   Undescended Testis.   Congenital Hernia.

A portion of the anterior abdominal wall seen from behind to shew the left testis, which is considerably smaller than its fellow, lying at the internal abdominal ring. A blue rod has been passed down the inguinal canal, and on the reverse of the specimen is seen to enter a short processus vaginalis. On the right side the tunica vaginalis, which also freely communicates with the peritoneal cavity, contains a well-developed testis.

Benjamin R., æt. 19, was admitted under Mr. Cooper Forster with symptoms of peritonitis associated with a strangu-

lated congenital hernia on the right side. Herniotomy was
performed, and the patient died two days later. At the autopsy
it was found that the left testis could be pushed through the
inguinal canal but immediately returned to its position at the
internal ring. *See Insp.* 1863, No. 254 ; and *Journal of Anat. &
Phys.* vol. xxii. p. 520.

2074   Undescended Testis.

The lower part of the anterior abdominal wall seen
from behind and mounted to shew a small undeveloped
testis situated at each internal ring. Rods passed
through the rings enter two short processes of the
peritoneum.

2075   Undescended Testis.

A left testis which in the recent state lay in the groin.
A dissection has been made to shew that the testis has
not descended further than the upper part of the tunica
vaginalis, and that this sac is continuous with an
unobliterated funicular process. The closure at the
internal ring is complete.

2076   Cyst with retained Testis.

A thin-walled cyst about as large as a pigeon's egg
which was connected with an undescended testis.

George E., æt. 14, was admitted under Mr. Howse in 1895
for a swelling in the right inguinal region. It was translucent and
irreducible. The right testis was not felt in the scrotum. The
swelling was exposed and found to consist of an undescended
testis together with the cyst which forms the preparation. The
cyst was situated within the cavity of the tunica vaginalis, above
and in front of the testis. It was excised and the testis, after
being freed from its attachment, was placed in the scrotum. *See
Surgical Reps.* vol. 176, Case 113.

2077   Undescended Testis.   Strangulation.

A small testis with its epididymis, both uniformly
blackened by extravasated blood. Histological exami-

nation shews the existence of many red blood corpuscles within and between the tubules.

Benjamin P., æt. 19, was admitted under Mr. Howse in 1879, having four days previously been attacked with sudden pain in the left groin. He was found to have an indistinct swelling with tenderness in that situation. An exploratory operation was performed, and the testis was removed. No communication between the tunica vaginalis and the abdominal cavity was found. The patient was stated to have always worn a truss for " hernia " on that side. *See Surgical Reps.* vol. 79, Case 107.

2078   Strangulation of the Testis.

A testis with the lower end of the spermatic cord mounted to shew the gland blackened by extravasated blood, and the cord closely twisted upon itself.

William M., æt. 20, was admitted under Mr. Lucas in 1891 for a painful swelling in the right groin, which had appeared suddenly three weeks previously. The right testis could not be felt in the scrotum, and the patient stated that it had never occupied its normal position. The swelling was explored and found to consist of the strangulated testis. The organ was removed and the patient was discharged well three weeks after the operation. *See Surgical Reps.* vol. 154, Case 191.

2079   Strangulation of the Testis.

A testis with a portion of the spermatic cord, removed by operation, and mounted to shew the gland and the cord swollen and engorged with blood. The body of the testis is blackened.

John M., æt. 27, was admitted under Mr. Howse in 1892 for pain in the left testicle, which came on suddenly while he was at work on the preceding day. On admission the testis and cord were found to be greatly swollen and very tender. The lymphatic glands in the groin were enlarged. An incision was made into the tunica vaginalis, and it was found that the testis had become twisted upon itself in the axis of the cord "three times round." The organ was removed and the patient was discharged well five weeks after the operation. *See Surgical Reps.* vol. 158, Case 137.

2080   **Suppuration of the Epididymis.**

The half of a right testis, shewing the epididymis to be much enlarged and filled with a soft curdy material somewhat resembling tuberculous deposit. Histological examination shews a condition of diffuse suppuration without caseation or giant-cells.

> From John P., æt. 30, who died from pyæmia following a carbuncle, with suppuration in the joints, lungs, and cerebral sinuses. The left epididymis was similarly affected to the right.

2081   **Abscess of the Epididymis.**

A testis laid open from behind to shew the epididymis greatly enlarged and thickened by inflammatory material. At the lower part is seen a small cavity, which in the recent state contained pus. On the reverse of the specimen is displayed the sac of a small hydrocele.

> George G., æt. 55, was admitted under Dr. Taylor for bronchitis and emphysema, associated with stricture of the urethra. He died seven weeks after admission, having suffered for the last sixteen days from rigors and a painful swelling of the epididymis. At the autopsy the kidneys were found to be hydronephrotic, and there was pyelonephritis. *See Insp.* 1889, No. 21.

2082   **Chronic Abscess of the Epididymis.**

A testis divided to shew a thick-walled abscess-cavity in connection with the epididymis. The cavity, which measures two and a half inches in length, contains a quantity of curdy material, and has at its lower end a short sinus opening externally through the skin of the scrotum. The wall of the cavity, which in parts is nearly half an inch thick, presents on histological examination a fibrous structure with no evidence of tubercle. The body of the testis is normal.

> John H., æt. 29, was under Mr. Golding-Bird's care in 1880 with an abscess in the scrotum which had been discharging curdy

pus for three weeks. Six years previously the patient contracted gonorrhœa, and a year later the scrotum became enlarged. This caused little inconvenience till eight weeks before admission, when he was attacked with violent pain and rigors.

2083   Calcareous Epididymis.

A testis with the lower end of the spermatic cord divided to shew in the situation of the epididymis an ovoid cavity containing pultaceous material, and having a thick fibrous wall, which has undergone partial calcification. The tunica vaginalis is thickened and adherent.

2084   Bony Deposit in the Testis.

A testis laid open to shew a mass of true bone situated in the body and epididymis of the organ, and measuring nearly an inch in its longest diameter. The bone has a compact structure and nodular surface.

William L., æt. 64, was admitted under Dr. Pavy in a moribund condition, having suffered for many years from difficulty in micturition. At the autopsy a stricture was found in the urethra, and the kidneys were in a condition of ascending nephritis. The liver was cirrhotic, and both the testes, which were atrophied, contained bony deposits. *See Insp.* 1883, No. 270.

2085   Hernia Testis.

A testicle which protruded through the skin of the scrotum, and was removed by operation. The protruding portion is seen to be covered with cauliflower-like granulations. Histological examination of the testis shews that the granulation tissue is implanted upon a fibroid testis.

John B., æt. 57, was admitted under Mr. Cock in 1855 with a hernia testis. The organ had been enlarged for about three months, and had been poulticed and lanced. Castration was performed and the patient made a good recovery.

U 2

2086   Hernia Testis.

A testis with a part of the scrotum divided to shew a
hernial protrusion of a portion of the gland through an
ulcerated opening in the skin. Histological examination
shews the tissue of the testis to be affected by recent
and chronic inflammation, dependent on syphilis.

2087   Syphilitic Orchitis.

A testis, measuring three and a half inches in its longest
diameter, laid open to shew the appearances of syphilitic
orchitis. The parenchyma of the organ is almost
entirely replaced by a dense fibrous material, in which
are embedded yellow masses of gummatous deposit. A
section through the globus major shews that a similar
change has affected the epididymis. On the reverse side
of the specimen the surface of the organ is seen to be
somewhat nodular ; the tunica vaginalis is neither
thickened nor adherent.

James B., æt. 77, was admitted under Mr. Lucas in 1894 for
a painless swelling of the left testis, which had been slowly
growing for four months. He had had syphilis 50 years before,
and had suffered from no illness of any kind since. The testis was
removed by operation, and the patient was discharged well one
month after the operation. See *Surgical Reps.* vol. 172, Case 219.

2088   Syphilitic Orchitis.

A pair of testes affected with chronic syphilitic inflammation. In the smaller preparation the tunica vaginalis
is considerably thickened, and is universally adherent,
while the cut section of the gland shews it to be converted into a firm mass of yellow material interspersed
with whiter fibrous bands. In the other, the tunica
vaginalis is partially adherent, the unobliterated portion
of the sac being distended in the recent state with
fluid. At the upper end of the body of the testis, which
is similarly affected to its fellow, there is a defined

gummatous nodule. Histologically the testicular stroma is infiltrated with small round cells, and presents an excess of fibrous tissue.

Presented by Mr. Jacobson, 1896.

2089 Syphilitic Orchitis.

A testis enlarged so as to measure four inches in length, and divided to shew its structure infiltrated by a deposit in parts caseous, in others fibroid. There are several loculi within the partially-obliterated sac of the tunica vaginalis. Histologically the deposit is syphilitic.

Charles H., æt. 37, was admitted under Dr. Davies-Colley in 1891 for a swelling on the right side of the scrotum, which had been noticed for one year. The testis was excised, a good deal of fluid escaping from the tunica vaginalis at the operation. The patient was discharged well five weeks after the operation. *See Surgical Reps.* vol. 153, Case 66.

2090 Gummatous Testis.

A testis considerably enlarged and divided to shew its substance entirely replaced by nodular masses of firm yellow material, with interlacing bands of fibrous tissue. The tunica vaginalis is adherent, and on the reverse of the specimen several sinuses are apparent in the skin of the scrotum.

Joseph S., æt. 44, was admitted under Mr. Howse in 1894 for a painless swelling of the right testis, associated with discharging sinuses in the scrotum. There was a history of gonorrhœa five years previously, but none of syphilis. Some months before admission a swelling appeared in the right clavicle and subsided under treatment. The testis was removed and the patient was discharged well one month after the operation. *See Surgical Reps.* vol. 170, Case 147.

2091 Gummatous Testis.

A testis enlarged so as to measure three and a half inches in length. It has been laid open to shew the body of the organ converted into a mass of yellow gummatous material intersected by coarse white bands of fibrous tissue. The tunica vaginalis is thickened and adherent.

Thomas W., æt. 31, was admitted under Mr. Birkett in 1857 with a swelling of the right testis, which had been increasing for four years. The organ was removed, and the patient made a good recovery.

2092   **Gummatous Testis. Fungating Ulcer of Scrotum.**

A testis with a portion of the scrotum, to which it is firmly adherent. The skin presents several irregular ulcers, in the base of which are masses of granulation tissue arising from the tunica vaginalis. On the reverse of the specimen a section through the organ displays a fibro-gummatous deposit in the body, the epididymis and vasa deferens being unaffected.

2093   **Gumma of the Testis.**

A section through an injected testis, shewing the gland to be converted into a mass of yellow material, the central parts of which have undergone necrotic cavitation. The tunica vaginalis is much thickened and adherent at the lower part. Above is seen the sac of a small hydrocele. Histologically the vessels present the appearance of syphilitic thickening, and the cheesy material is seen to be gummatous.

Presented by Mr. Aston Key.

2094   **Gummata of the Testis.**

The half of a left testis, the cut surface of which shews several gummatous nodules projecting from the fibrous body of the organ. The tunica vaginalis is closely adherent, much thickened, and is of almost cartilaginous hardness. On the reverse of the specimen there is an area of granulation tissue which formed the base of an ulcer of the scrotum.

Harry K., æt. 40, was admitted under Mr. Golding-Bird in 1895 for an enlargement of the left testis associated with ulceration of the scrotum. He had contracted syphilis three years previously, and the swelling of the testis had been observed for two years. The scrotum was tapped, and clear fluid drawn off on four occasions. The ulcer appeared on the scrotum three weeks before admission. The testis was excised, and the patient was discharged relieved three weeks later. *See Surgical Reps.* vol. 179, Case 251.

2095   Calcareous Gumma of the Testis.

A testis of normal size, laid open to shew at its upper part a rough calcareous nodule measuring rather more than half an inch in diameter. The rest of the organ is converted into dense fibrous tissue.

William J., æt. 57, was admitted under Mr. Howse for an epitheliomatous stricture of the œsophagus, and died on the following day from broncho-pneumonia. There were scars upon his wrist and forehead. *See Insp.* 1891, No. 316.

2096   Tuberculous Testis.

The half of an injected testis, shewing the body of the organ uniformly infiltrated with yellowish material. The sac of the tunica vaginalis is obliterated, and its parietal layer, which in the recent state was exposed by ulceration of the scrotum, is greatly thickened and covered by granulation tissue. The epididymis is much enlarged by caseous deposit.

From a man, æt. 23, who had suffered from a painful enlargement of the right testis. Some months after the onset of symptoms suppuration supervened, an abscess opened externally, and the body of the organ was extruded from the scrotum. Mr Bryant removed the testis with a portion of the spermatic cord, and the patient made a good recovery.

Presented by Mr. Bryant, 1892.

2097   Tuberculous Testis.

A testis with a portion of the scrotum and spermatic cord removed by operation. The testis is somewhat enlarged and contains miliary tubercles. It lies surrounded by a softening caseous mass, apparently consisting of tuberculous deposit in the epididymis and in the sac of the tunica vaginalis.

John H., æt. 38, was admitted under Mr. Davies-Colley in 1896 for a painful swelling of the left testis, which had been noticed for three months. Fluctuation was detected, and there was a small sinus over the situation of the head of the epididymis. The testis was excised, and the patient discharged well seven weeks after the operation. *See Surgical Reps.* vol. 183, Case 186.

2098   Tuberculous Testis.   Ulceration of Scrotum.

The half of a testis with a portion of the scrotum and the spermatic cord. The organ is enlarged, measuring three inches in its longest diameter, and is uniformly infiltrated with a yellowish deposit having the histological characters of caseous tubercle. The tunica vaginalis is adherent, and there is a large ulcer destroying the integument and exposing the testis. The vas deferens is also thickened by caseous deposit.

2099   Tuberculous Testis.

A right testis divided to shew its epididymis converted into a soft cheesy material, and its body thickly beset with small yellow nodules. Histologically the deposit is seen to consist of caseating tubercle.

George G., æt. 36, was admitted under Dr. Gull for phthisis, from which two months later he died. At the autopsy miliary tubercles were found in the liver and kidneys. The vesiculæ seminales were scrofulous. *See Insp.* 1864, No. 64; and *Preps.* 2069 and 2144.

2100   Tuberculous Abscess of the Testis.

A testis laid open to shew at its upper part the cavity of a small abscess with caseous walls. The epididymis has a fibro-caseous structure, and there are small tuberculous nodules in the body of the testis.

George N., æt. 38, was admitted under Dr. Barlow for phthisis, from which five weeks later he died. At the autopsy the liver was found to be fatty, and the kidneys contained numerous tubercles. *See Insp.* 1857, No. 210.

2101   Tuberculous Testis.

An infant's testis considerably enlarged and laid open to shew that its proper structure is replaced by a fairly uniform yellow deposit having the histological characters of caseous tubercle.

William M., æt. 2½, was admitted under Dr. Bryant for an enlargement of the right testis which had been noticed for 3 or 4 months. An abscess formed and the gland was removed. Four weeks later the patient died, and at the autopsy miliary tubercles were found on the lungs and the peritoneal cavity was obliterated by adhesions. *See Insp.* 1858, No. 224; and *Drawing*, No. 415 (30).

2102   Tuberculous Testis.

A testis divided to shew its epididymis enlarged and filled with caseous material. In the body of the testis are seen numerous miliary tubercles more closely aggregated at its attachment to the epididymis.

George C., æt. 32, was admitted uuder Mr. France for a melanotic growth of the left eyeball. Three months later he died and at the autopsy the growth was found to have invaded the dura mater, and there were numerous secondary deposits. Tuberculous deposit was found in the testis, and in each vas deferens, but in no other part of the body. *See Insp.* 1859, No. 119.

2103   Tuberculous Epididymis.

An injected testis divided to shew the epididymis converted into a disintegrating caseous mass as large as the body of the organ. In the substance of the testis there is a tuberculous nodule about half an inch in length. The tunica vaginalis is thickened and adherent.

2104   **Squamous-celled Epithelioma of the Testis.**

A testis with its coverings divided to shew the organ uniformly infiltrated with a soft growth and enlarged so as to measure 3 inches in its longest diameter.   On the reverse of the specimen the skin of the scrotum presents a rounded ulcer about the size of a florin, with hard edges and a base covered with prominent granulations. The growth in the scrotum has invaded and infiltrated the testis, in which it has the histological characters of a squamous-celled epithelioma, presenting cell-nests and areas of cystic degeneration.

From a patient whose testicle was removed by Mr. Birkett.

2105   **Carcinoma of the Testis.**

The half of a testis measuring six inches in its longest diameter and having an irregular lobulated exterior. The cut surface shews the proper glandular tissue to be replaced by a new growth, in parts soft and necrotic, in others harder and more fibrous.   Histologically the deposit has the characters of a spheroidal-celled carcinoma.

From a patient whose testis was removed by Mr. Aston Key.

2106   **Carcinoma of the Testis.**

A testis somewhat enlarged and laid open to shew its substance occupied by nodular masses of soft vascular growth separated from each other by strands of firmer white tissue.   The tunica vaginalis is thickened and adherent.   Histologically the growth is a spheroidal-celled carcinoma.

Godfried W., æt. 19, was admitted under Mr. Durham for growth of the testis.   The organ was removed and a few days later the patent died.   At the autopsy secondary deposits were found in the liver and lumbar glands. *See Insp.* 1862, No. 6.

2107 **Carcinoma of the Testis.**

A left testis measuring four inches in length and uniformly infiltrated by a firm white deposit. The cut section shews numerous interlacing strands of fibrous tissue. Histologically the growth is a spheroidal-celled carcinoma.

Richard S., æt. 40, was admitted under Mr. Cooper Forster in 1858 with a hard nodulated enlargement of the left testis which had been gradually increasing in size for the last three years. The organ was removed.

2108 **Carcinoma of the Testis.**

The half of a testis having a nodular exterior, and enlarged so as to measure three inches in length. It is uniformly infiltrated by a growth, in parts fibrous and in others of a softer consistency. Histologically the deposit is a spheroidal-celled carcinoma.

From a man, æt. 64, who made a good recovery after the removal of the tumour.

2109 **Carcinoma of the Testis.**

A testis partially divided to shew at its upper end a white encapsuled growth measuring an inch in diameter. Histologically it is a spheroidal-celled carcinoma with scanty stroma.

Presented by Mr. Jacobson, 1896.

2110 **Carcinoma of the Testis.**

A testis injected and divided by sagittal section. The organ is enlarged so as to measure 3 inches in diameter, and is firmly adherent to the tunica vaginalis, the sac of which is obliterated except at its upper end. The epididymis and body of the testis are both uniformly infiltrated by a spheroidal-celled carcinoma with scanty stroma.

John C., æt. 38, was admitted under Mr. Jacobson in 1893 for a painless, nodular swelling of the left testis, which had been first noticed 3 years previously. The testis was excised and the patient was discharged well five weeks after the operation. *See Surgical Reps.* vol. 167, Case 23.

2111   Carcinoma of the Testis.

The half of an injected testis, about twice the normal size, and presenting on its cut surface numerous small cysts. Histologically it is found to be infiltrated with a deposit of carcinoma, some of the alveoli of which are lined by cylindrical epithelium. The surface of the organ is nodulated.

2112   Small Round-celled Sarcoma of the Epididymis.

The testes of a youth mounted to shew the epididymis of each enlarged by a uniform deposit of growth and a similar deposit in the cord of the left. The testes themselves are unaffected. Histologically the growth is a small round-celled sarcoma.

Samuel S., æt. 16, was admitted under Dr. Rees for paraplegia of about 5 weeks' duration. A fortnight later he died, and at the autopsy malignant growth was found in the spine, jejunum, heart, kidneys, and lymphatic glands. *See Insp.* 1862, No. 1; and *Prep.* 934.

2113   Sarcoma of an undescended Testis.

A portion of an undescended right testis which weighed 36 ounces, and was found free from adhesions in the peritoneal cavity. It is uniformly infiltrated by a soft white deposit having the characters of a small round-celled sarcoma. On the reverse of the specimen the epididymis is seen to be similarly infiltrated.

Richard R., æt. 52, was admitted under Dr. Moxon with

pain in the back associated with a tumour occupying the position of a distended bladder. Four months later he died and at the autopsy secondary deposits were found in the vertebræ, thymus, and in the mediastinal and abdominal lymphatic glands. *See Insp.* 1882, No. 390.

2114   Sarcoma of the Testis.

A testis enlarged and infiltrated by growth which has undergone degenerative softening so that the organ is converted into a large cyst-like cavity whose walls are formed of the tunica albuginea. The cavity measures six by three and a half inches, and its internal surface is rough and in parts calcareous. In the recent state it contained 14 ounces of brown mucoid fluid in which under the microscope were seen cholesterine crystals and ciliated epithelial cells. At the lower and anterior part of the preparation are seen the remains of the gland, which histologically is infiltrated with round-celled sarcoma. The vas deferens, coloured red, is normal.

John C., æt. 72, was admitted under Mr. Bryant for an enlargement of the scrotum of many years' duration, and a tumour in the right groin, first noticed three months before admission. He died four months later, and at the autopsy a large sarcomatous growth was found extending from the right iliac fossa downwards into the thigh. *See Insp.* 1888, No. 197.

2115   Round-celled Sarcoma of the Testis.

A testis somewhat enlarged and laid open to shew at its lower part an encapsuled mass of growth of the shape and size of the normal organ. The growth, which is soft and hæmorrhagic, is dotted over with numerous small cysts. Histologically the malignant deposit has the characters of a round-celled sarcoma.

John A., æt. 47, was admitted under Mr. Durham in 1893 for a painful swelling of the left testis of five months' duration. The testis was excised and the patient was discharged well twelve days later. *See Surgical Reps.* vol. 163, Case 107.

2116   Round-celled Sarcoma of the Testis.

A testis uniformly enlarged so as to measure four inches in length, and occupied by a homogeneous growth having the histological characters of a round-celled sarcoma. The surface of the organ is smooth and the tunica vaginalis is adherent.

From George J., æt. 40, who noticed that his testis had been enlarging for five years before its removal by Sir Astley Cooper in 1807. The enlargement was attributed to an injury.

2117   Round-celled Sarcoma of the Testis.

A testis which measures four inches in length and is uniformly infiltrated by a white growth in which are visible a few nodules of cartilage. At the upper part of the organ the deposit has undergone degenerative softening. The tunica vaginalis is thickened and adherent. Histologically the growth is a round-celled sarcoma.

From a man, æt. 48, whose testicle was removed by Mr. Birkett in 1864. He made a good recovery. *See Drawing,* 417 (10).

2118   Round-celled Sarcoma of the Testis.

A testis measuring six inches in its longest diameter and laid open to shew at its lower part a homogeneous infiltrating growth in which are embedded a few minute nodules of hyaline cartilage. At the upper part the growth is softer and presents numerous ragged cavities, some of which contain blood-clot. The exterior of the testis is smooth and the tunica vaginalis is thickened and inflamed. Histologically the growth is a round-celled sarcoma.

Richard F., æt. 25, was admitted under Mr. Birkett in 1860 for an enlargement of the right testicle which had been noticed for about three years. Three months previously the tumour was punctured, after which it rapidly increased in size, and on admission it was found to be pulsating. The testis was removed and the patient recovered. *See Drawings,* 416 (65 & 66).

2119   Round-celled Sarcoma of the Testis.

An injected testis measuring four and a half inches in length and infiltrated by a growth which forms bossy prominences on the exterior of the organ. Histologically the growth has the characters of a round-celled sarcoma.

John S., æt. 30, was admitted under Mr. Bransby Cooper in 1828 with a tumour of the testis. He was transferred to the care of Dr. Bright with hemiplegia and pulmonary symptoms. Thirteen days after admission he died, comatose, and at the autopsy secondary deposits were found in the brain and in the lungs. *See Insp.* vol. 14, p. 15; and *Prep.* 1576 (64) [2nd Edit.].

2120   Sarcoma of the Testis.

An injected testis uniformly enlarged so as to measure three and a half inches in its longest diameter. The gland is infiltrated with growth shewing areas of softening at its lower part and a few scattered cysts. The surface is smooth, the epididymis normal, and the cord not thickened. Histologically the growth is composed of large, round, and very long spindle-cells, and is in parts hæmorrhagic.

From a man, æt. 30, who had noticed a gradual enlargement of his left testis for about nine months. He was in good health a year after the removal of the organ.

Presented by Dr. Stevens.

2121   Sarcoma of the Testis.

A testis measuring three and a half inches in length, laid open to shew its tissue uniformly infiltrated with a hæmorrhagic growth in which are numerous small cysts. The surface of the organ is smooth and free from adhesion. Histologically the growth is a sarcoma with round and spindle cells.

John S., æt. 26, was admitted under Mr. Cock in 1858, with a painless swelling of the right testis which had been noticed about two years.

2122   Sarcoma of the Testis.

A portion of a large mass of growth which occupied the scrotum and inguinal regions of a child two years of age. Histologically it presents the characters of spindle-celled sarcoma. *See Wax Models, 53 & 54.*

Presented by Mr. Montague Gosset.

2123   Spindle-celled Sarcoma of the Testis.

A testis measuring five inches in length and uniformly infiltrated by growth which forms nodular prominences upon its surface. The cut section shews the deposit to be in many parts darkened by extravasated blood. The coverings of the organ are greatly thickened and the cavity of the tunica vaginalis is obliterated. Histologically the growth is a spindle-celled sarcoma.

Presented by Mr. Symonds, 1893.

2124   Alveolar Sarcoma of the Testis.

A testis somewhat enlarged and partially infiltrated by a white deposit seen on section to consist of soft nodules of various sizes, separated from each other by strands of a fibrous-looking material. Histologically the growth is a large round-celled sarcoma, the stroma of which presents an alveolar arrangement. The epididymis appears to be normal.

Presented by Mr. Golding-Bird.

2125   Alveolar Sarcoma of the Testis.

A testis enlarged so as to measure three inches in length and laid open to shew the gland occupied by closely aggregated nodules varying in size from a line to half an inch in diameter. The nodules are of soft consistency, and having undergone less contraction from the effects of the alcohol in which the preparation was hardened than has the testicular stroma, they form prominent rounded elevations. Histologically the growth is a large round-celled sarcoma with alveolated stroma.

Robert F., æt. 26, was admitted under Mr. Golding-Bird in 1894 for a swelling of the right testis which had been noticed seven months. It was tender on manipulation, and there was some thickening of the cord. The testis was excised and the patient was discharged in good health three weeks later. *See Surgical Reps.* vol. 173, Case 169.

2126 Alveolar Sarcoma of the Testis.

A testis enlarged so as to measure three inches in length, and uniformly infiltrated with an almost homogeneous deposit of soft malignant growth. The exterior of the organ is smooth and free from adhesions. The epididymis appears to be normal. Histologically the growth is a large round-celled sarcoma with alveolated stroma. *See Preps.* 1523, 1527 [2nd Edit.].

Presented by Sir Astley Cooper.

2127 Alveolar Sarcoma of the Testis.

A portion of a testis which weighed 12 ounces, and is infiltrated by a soft white growth having the histological characters of a large round-celled sarcoma with alveolated stroma. The lower portion of the tumour has undergone necrotic softening. The tunica vaginalis is adherent.

William R., æt. 48, was admitted under Mr. Symonds in 1886, with a swelling of the testis which had been noticed for twelve months. The organ was removed and some months later the growth recurred in the unguinal and lumbar glands. The patient died about seven months after the operation. *See Surgical Reps.* vol. 124, Case 11.

2128 Alveolar Sarcoma of the Testis.

A partially injected testis measuring four and a half inches in length, laid open to shew the organ infiltrated by a deposit of growth which presents numerous areas of necrotic softening. The tunica vaginalis is adherent except at the upper part, where its serous lining is coated with lymph. Histologically the growth is a large round-celled sarcoma with alveolated stroma.

Jonathan K., æt. 28, was admitted under Mr. Aston Key in 1828, with a painless enlargement of the right testis which had been noticed for five years. The tumour was tapped, and afterwards the patient became feverish and developed a scarlatiniform rash. The scrotum became gangrenous, and the patient died six days after the operation. *See Insp.* vol. 5, p. 150.

2129   Alveolar Sarcoma of the Testis.

The half of a testis measuring four inches in length, the glandular structure of which is replaced by a deposit of growth which is in part necrotic. The surface of the organ is somewhat irregular and the tunica vaginalis is adherent. Histologically the growth is a large round-celled sarcoma, the stroma of which presents an alveolar arrangement.

Thomas T., æt. 56, was admitted under Mr. Hilton in 1855 for a swelling of the right testis, which had been gradually increasing for twelve months. *See Drawing,* 419 (57).

2130   Myxo-Sarcoma of the Testis.

The half of child's testis measuring two and a half inches in length and infiltrated with a white gelatinous growth which in the recent state presented a few areas of extravasated blood. The organ has a nodulated surface, and histologically the tumour is a spindle-celled sarcoma with myxomatous degeneration. .

From a child, æt. 6, whose mother had noticed a rapidly increasing swelling of the testis for five months before its removal.

Presented by Dr. Moir, 1897.

2131   Myxo-Sarcoma of the Testis.

A testis laid open to shew the organ occupied by partially encapsuled nodular masses of growth in which are seen numerous cysts. Histologically the deposit is a vascular round-celled sarcoma with myxomatous change. *Presented by Mr. Aston Key.*

2132   Cystic Chondro-Carcinoma of the Testis.

A testis four inches in length, laid open to shew its structure replaced by a growth of varied character and consistency. At the upper part it is tolerably firm and homogeneous. Separated from this by a capsule are masses of soft hæmorrhagic material surrounded by a fibrous substance in which are seen numerous translucent nodules of hyaline cartilage and a few cysts. Histologically the cysts are lined with stratified epithelium, and the softer material is seen to have the characters of spheroidal-celled carcinoma.

Presented by Mr. Cock, 1866.

2133   Chondro-Sarcoma of the Testis.

A testis which measures three and a half inches in length, and is for the most part converted into a translucent grey material traversed by strands of a whiter substance. Histological examination shews the deposit to consist of hyaline cartilage in a matrix of sarcomatous spindle cells.

2134   Cystic Chondro-Sarcoma of the Testis.

A testis enlarged so as to measure four inches in its longest diameter and infiltrated with growth. The peripheral portions of the tumour consist of a homogeneous material with a few scattered cysts. The central area, in which the structure is less dense and the cysts more numerous, is partially surrounded by a fibrous capsule, beneath which there is considerable degenerative softening. The tunica vaginalis is adherent and greatly thickened. Histologically the growth is a sarcoma containing cartilage.

x 2

2135   Cystic Chondro-Sarcoma of the Testis.

A testis measuring two and three quarter inches in diameter, the tissue of which is entirely replaced by a growth which on section presents a convoluted structure and contains numerous cysts. The tunica vaginalis is adherent, except over the upper part of the organ. Histologically the growth is a spindle-celled sarcoma containing hyaline cartilage.

Alfred C., æt. 29, was admitted under Mr. Symonds in 1890 for an enlargement of the left testis, which had been first noticed two months previously. The testis was excised, and the patient was discharged well one month later. *See Surgical Reps.* vol. 150, Case 69.

2136   Cystic Chondro-Sarcoma of the Testis.

A testis enlarged so as to measure three inches in length, divided to shew the entire organ infiltrated by a firm growth. The cut section exhibits interlacing bands of white fibrous tissue with numerous translucent areas of cartilage and a few small cysts. On the reverse of the specimen the capsule is seen to be somewhat thickened, and several cysts are visible beneath it. Histologically the structure is that of a fibro-sarcoma with hyaline cartilage.

William B., æt. 30, was admitted under Mr. Cock in 1862 with a painless enlargement of the testis, which had been observed for six months. His general health was unaffected.

2137   Myo-Sarcoma of the Testis.

The half of an infant's testis measuring an inch and a half in length and uniformly infiltrated by a soft white growth, which histologically has the structure of adeno-sarcoma with muscular tissue.

From a child, æt. 22 months, whose testis was noticed to be enlarged at birth. The tumour increased rapidly during the three months preceding its removal by Mr. Cooper Forster in 1864.

2138   Rhabdo-Sarcoma of the Testis.

A testis measuring three inches in length and infiltrated by a growth in parts homogeneous and in parts cystic. The cysts have a smooth lining and vary in size from a pin's head to half an inch in diameter. The tunica vaginalis is thickened and adherent. Histologically the growth consists of connective tissue, traversed by striped muscular fibres, and containing a few scattered nodules of cartilage.

Removed by Sir Astley Cooper and regarded by him as a specimen of "tubular" disease of the testis, the cysts being supposed to arise from dilatation of the seminiferous tubes.

2139   Cystic Myo-Sarcoma of the Testis.

A testis measuring five inches in length, and presenting on its cut surface numerous cysts, some of which are filled with pultaceous material. The matrix in which the cysts are embedded is formed by a white dense tissue which histologically has the characters of a myo-sarcoma. The tunica vaginalis is adherent and considerably thickened.

From Joseph W., æt. 44, who had noticed a painless enlargement of the testis for some months. When removed the organ weighed two pounds.

Presented by Mr. Hilton.

2140   Myo-Chondro-Sarcoma of the Testis.

A testis enlarged so as to measure four inches in length, and entirely occupied by a soft growth. The cut section shews the deposit to be in parts hæmorrhagic and to present a few cysts and areas of necrotic softening. Histologically the growth is a round-celled sarcoma containing unstriped muscular fibres and numerous islets of cartilage. The tunica vaginalis is thickened and partially adherent.

Removed by Mr. Dodd from a young epileptic patient.

2141   Cystic Disease of the Testis.

A testis, having a nodulated surface and measuring three and a half inches in length. It has been laid open to shew that it is occupied by numerous cysts varying in size from a line to three quarters of an inch in diameter, separated from each other by a fibrous stroma containing fragments of cartilage. The cysts have a smooth lining, and many of them contain a pultaceous material. Histological examination shews that they are lined with ciliated columnar epithelium, and that the matrix is fibrous and contains hyaline cartilage.

From a man, æt. 22, whose testis was removed by Mr. Cock in 1861. The enlargement was first noticed when the patient was five years of age. It was tapped on two occasions and a small quantity of mucoid fluid evacuated. *See Drawing*, 416 (60).

2142   Cystic Disease of the Epididymis.

A testis with the epididymis and the spermatic cord mounted to shew the epididymis converted into a congeries of cysts, having thin translucent walls and varying in size from a third of an inch to two inches in diameter. The whole epididymis is more than four times as large as the testis. The cysts have a smooth lining, they do not communicate with each other, and in the recent state they were filled with an opalescent fluid containing spermatozoa.

Samuel W., æt. 65, was admitted under Dr. Goodhart for chronic Bright's disease, and died from uræmia. At the autopsy cysts were found in the other epididymis, and histological examination of the more solid part shewed some of the tubules to be slightly dilated. *See Insp.* 1890, No. 57.

2143   Cyst of the Epididymis.

A testis with the epididymis, in the globus major of which is seen a thin-walled cyst measuring about half an inch in diameter.

2144   **Cyst of the Epididymis.**

A left testis with the epididymis mounted to shew in the globus major a thin-walled cyst measuring about a third of an inch in diameter, which in the recent state contained turbid fluid.

George G., æt. 36, was admitted under Dr. Gull for phthisis, from which two months later he died. At the autopsy miliary tubercles were found in the liver and kidneys. The right testis and the vesiculæ seminales were scrofulous. *See Insp.* 1864, No. 64 ; and *Preps.* 2069 & 2099.

2145   **Cyst of the Testis.**

A testis divided to shew in its substance towards the globus major of the epididymis a smooth-walled cyst measuring half an inch in diameter, which in the recent state contained clear fluid. Histologically the structure of the testis is normal.

John B., æt. 73, was admitted under Dr. Shaw for Bright's disease and died from cardiac failure. *See Insp.* 1891, No. 53.

Section XXXVIII.—DISEASES OF THE TUNICA VAGINALIS.

2146   Patent Tunica Vaginalis.

The testis of a child with its coverings and the spermatic cord. The tunica vaginalis has been laid open to shew that its cavity is continuous with that of the peritoneum.

2147   Patent Tunica Vaginalis.

An adult testis enclosed in the tunica vaginalis from which a rod has passed upwards along the spermatic cord to demonstrate the persistence of a communication between the sac and the general peritoneal cavity.

2148   Hydrocele.

A vertical section through a right testis and its tunics shewing a small vaginal hydrocele, and above it and completely separated from it the lower end of the sac of an inguinal hernia.

David B., æt. 67, was admitted under Mr. Jacobson in 1894 for an inguinal hernia on the right side, associated with a small hydrocele. An operation for the radical cure of the hernia was performed, and at the same time the testis was excised. The patient was discharged well two months after the operation. *See Surgical Reps.* vol. 174, Case 57.

2149   Double Hydrocele.

A pair of testes with their coverings mounted to shew the condition of the parts after treatment for the relief of double vaginal hydrocele. On the right side the sac is entirely obliterated, and the two layers of the tunica vaginalis are coherent and thickened. On the left the sac, which measures three inches in its longest diameter, is coated with a thick deposit of inflammatory lymph.

John M., æt. 64, was admitted under Mr. Birkett with symptoms of urinary disease. Sixteen days later he died, and at the autopsy the prostate was found to be enlarged and the kidneys were hydronephrotic. The sac of the uncured hydrocele had been tapped and injected some days before his death. *See Insp.* 1859, No. 28.

2150   Suppurating Hydrocele.

A vertical section through a testis and its coverings with the spermatic cord. The preparation has been injected and the section shews the sac of the tunica vaginalis to be distended and filled with coagulated lymph. The parietal layer of the sac is much thickened and on its outer surface is seen an ulcer, apparently the result of tapping.

2151   Sloughing of a Hydrocele.

A globular fibrous sac about three inches in diameter, smooth within and somewhat rough externally, which is said to have separated from the scrotum by sloughing consecutive to an operation for the relief of a hydrocele.

Presented by Mr. Aston Key.

2152   Hydrocele with Adhesions.

The sac of a hydrocele partially injected and laid open to shew adhesions between the testis and the parietal layer of the tunica vaginalis.

2153   Hydrocele with Adhesions.

A testis with the tunica vaginalis laid open to shew the sac dilated so as to measure five inches in length. The testis lies at the lower and hinder part of the cavity and several claw-like processes of fibrous tissue proceed from it and two of a filamentous character are attached to the opposite wall of the sac.

2154   Hydrocele with Thickened Sac.

A vertical section through a testis and its tunica vaginalis with a portion of the skin of the scrotum. The tunica vaginalis has its parietal layer thickened and is lined by a deposit of organised lymph which on one side of the sac measures an inch and a half in thickness. The testis is seen at the lower part of the tumour and above it is a cavity which contains recent lymph.

Presented by Mr. J. Adamson.

2155   Chronic Hydrocele.

The sac of a hydrocele measuring five inches in length laid open to shew its walls to be thick and fibrous and the testis to be firmly adherent by its posterior half to the lower and hinder part of the sac. The hydrocele extends rather more than two inches along the spermatic cord.

William D., æt. 62, was admitted under Mr. Howse for injuries received by falling from a roof and died the following day. *See Insp.* 1892, No. 406.

2156   Hydrocele : Calcification of Sac.

A dry preparation of a hydrocele preserved to shew numerous white calcareous plates in the wall of the tunica vaginalis.

2157   Cured Hydrocele.

A testis with its tunics laid open to shew the layers of the tunica vaginalis thickened and adherent to each other except at the upper part of the organ, where the remains of a hydrocele are still apparent.

From a man, æt. 55, who died from Bright's disease.

2158   Hydrocele of the Spermatic Cord.

A portion of a spermatic cord dissected to shew in its course a small cyst the walls of which are thin, fibrous, and trabeculated.

2159   Hydrocele of the Spermatic Cord.

A tubular cyst measuring two inches in length and half an inch transversely which was situated in the spermatic cord. It has thin translucent walls and is divided into two nearly equal parts by a partial transverse septum. The fibres of the cremaster muscle may be seen upon the surface of the sac.

Alfred C., æt. 7, was admitted under Mr. Davies-Colley with a swelling extending from the external abdominal ring to the testis on the right side, which had been noticed for three months. The cyst, which in the recent state contained a clear straw-coloured fluid, was separated from the testis without difficulty. *See Surgical Reps.* 1887, Case 171.

2160   Hydrocele of the Spermatic Cord.

A testis with the spermatic cord dissected to shew immediately above the sac of the tunica vaginalis and completely separated from it a globular cyst extending three inches in the course of the cord. The walls are thick and fibrous and the internal surface somewhat roughened.

2161   Hydrocele of the Spermatic Cord.

A thin-walled globular cyst about an inch in diameter which was dissected from the structures forming the spermatic cord.

Charles J., æt. 3, was admitted under Mr. Davies-Colley in 1893 for an oval fluctuating swelling in the right groin which was found to be irreducible and translucent. The cyst was dissected from the surrounding tissues, and the patient was discharged well a fortnight later. *See Surgical Reps.* vol. 165, Case 286.

2162   Hydrocele of the Spermatic Cord.

A sagittal section through a right testis and the spermatic cord, shewing immediately above the testis, and completely shut off from the sac of the tunica vaginalis, a pyriform cyst measuring three inches in length. The wall of the cyst is tough and fibrous and its lining is rugose. The tunica vaginalis is adherent to the testis except on the upper part.

James H., æt. 56, was admitted under Mr. Davies-Colley, and died five days after the operation of lithotrity. The patient had worn a truss for some years over a swelling in the right inguinal region which on admission was found to be a hydrocele. *See Insp.* 1896, No. 443.

2163   Loose Body from a Hydrocele.

A smooth spherical body, measuring a third of an inch in diameter, and formed of dense fibrous tissue resembling cartilage. It was found loose in the sac of a hydrocele which presented evidences of inflammation.

2164   Loose Bodies from the Tunica Vaginalis.

Three specimens of calcareous deposit found loose in the sac of the tunica vaginalis. One is smooth and spherical, measuring a line in diameter, whilst the others, which are somewhat larger, are flattened and have an irregular shape.

2165   Cyst of the Tunica Vaginalis.

A thin-walled cyst measuring rather more than an inch
in diameter. The flattened and roughened surface of
the sac corresponds to its attachment to the parietal
layer of the tunica vaginalis. It is stated that the cyst
was connected neither with the testis nor the epididymis.

Joseph R., æt. 24, was admitted under Mr. Lucas in 1895
with a varicocele and a swelling in the course of the spermatic
cord. The cyst was dissected from the surrounding tissues and
the varicocele was excised. The patient was discharged well three
weeks after the operation. *See Surgical Reps.* vol. 178, Case 249.

2166   Calcareous Tunica Vaginalis.

A portion of a tunica vaginalis partially separated into
two sacs, the smaller of which contained the testis and
the larger the epididymis. The wall of the latter cavity
has undergone almost complete calcareous degeneration,
being rough, thick, and nodular.

2167   Acute Hæmatocele.

The half of the sac of a hæmatocele distended by laminated blood-clot. The tunica vaginalis is not thickened.

2168   Hæmatocele.

A vertical section through a testis and the sac of a
hæmatocele. The gland is seen to be situated at the
lower and hinder part of the cavity and to be somewhat
flattened. The walls of the hæmatocele are thick and
fibrous, and its internal surface is nodular and stained
by adherent blood-clot.

2169   Hæmatocele.

A section through a testis and the sac of a hæmatocele.
The testis, which is of small size, is seen embedded in
the fibrous wall of the hæmatocele, on the inner surface
of which is a thick deposit of recent lymph.

From John E., æt. 10, upon whom Mr. Hilton operated for hæmatocele in 1860. The patient was known to have had a scrotal swelling since the age of two years.

2170   Hæmatocele.

A testis with its coverings divided to shew immediately above the gland the sac of a hæmatocele measuring about two inches in diameter. The walls of the sac are greatly thickened and formed of dense fibrous tissue, the innermost layers being of cartilaginous consistency. On the interior of the cavity is a deposit of granular blood-clot.

From John L., æt. 53, who was operated upon for hæmatocele in 1869. The enlargement of the scrotum had been noticed for three years, and was not apparently attributable to injury.

2171   Hæmatocele.

A testis with its coverings and the lower end of the spermatic cord. The parietal layer of the tunica vaginalis is greatly thickened, and its sac distended so as to measure five inches in diameter. The interior of the sac is fibrous, rugose, and in parts calcified. The testis is seen at the hinder and lower part of the tumour.

Removed by Mr. Thomas Bryant in 1865.

2172   Hæmatocele.

A globular sac, at the back of which are seen the lower end of the spermatic cord and a portion of the skin of the scrotum. The sac, which measures three and a half inches in diameter, has thick fibrous walls and its interior is roughened and partially coated by a granular deposit of fibrin.

From a man, æt. 40, who was admitted under Mr. Poland in 1860 for a scrotal tumour on the right side, which had been noticed since childhood and during the two months preceding admission had rapidly increased in size. The tumour was removed after an ineffectual attempt to empty it by tapping. On examination, the hæmatocele was found to contain a grumous chocolate-coloured matter and cholesterine.

2173   Hæmatocele of the Spermatic Cord.

A testis with its coverings divided to shew at the lower end of the spermatic cord a fibrous-walled sac measuring two inches in length and partially filled with recent blood-clot. There is also a small vaginal hydrocele.

Patrick S., æt. 48, was admitted under Mr. Davies-Colley in 1894 for a painful swelling on the left side of his scrotum, which had been first noticed 7 months previously. The scrotum was tapped, and about 2 ounces of clear fluid withdrawn, after which a considerable, more solid, swelling remained. This, together with the testis, was excised, and was found to consist of the thick-walled sac filled partly with translucent gelatinous material and partly with a grumous fluid composed of blood and pus. *See Surg. Reps.* vol. 171, Case 210.

Section XXXIX.—INJURIES AND DISEASES OF THE SCROTUM.

Laceration : 2174.
Tuberculous Sinus : 2175.
Epithelioma : 2176–2180.

2174   Laceration of the Scrotum.

A scrotum shewing on either side of the median raphe a lacerated wound of the skin and tunica vaginalis, through which the testis protrudes.

William F. was admitted under Mr. Poland for the injuries shewn in the preparation, associated with fracture of the pelvis. The testes were replaced in the scrotum, and the patient died on the following day. *See Insp.* 1864, No. 67.

2175   Tuberculous Sinus in the Scrotum.

A right testis with the portion of the scrotum in contact with its lower extremity. There is a sinus in the scrotum communicating with a caseous abscess at the lower end of the epididymis. The epididymis itself is fibro-caseous.

William H. was admitted under Mr. Morgan in 1831 for retention of urine, and died from pyæmia. *See Insp.* vol. 15, p. 118.

2176   Epithelioma of the Scrotum.

A portion of a scrotum shewing a circular ulcer measuring an inch and three quarters in diameter and exposing in its base flattened nodules of malignant deposit. The edge of the ulcer is thin and somewhat overhanging. Histologically the growth is a squamous-celled epithelioma.   *Presented by Mr. Morgan.*

2177   Epithelioma of the Scrotum.

A portion of skin excised from the scrotum and mounted to shew an oval ulcer nearly an inch and a half in length, having a raised, somewhat everted edge and a base covered with prominent nodules of growth.   Histologically the deposit is a squamous-celled epithelioma. *See Wax Model, 91.*   *Presented by Mr. Aston Key.*

2178   Epithelioma of the Scrotum.

A portion of the skin of a scrotum shewing chimney-sweep's cancer.   In one part the growth forms a prominent warty tumour with overhanging edges and thick club-like excrescences.   At another there is a crater-like ulcer with crenated margin and malignant granulations on its sides.   Histologically the growth is a squamous-celled epithelioma with cell-nests.

2179   Epithelioma of the Scrotum.

A portion of the skin of a scrotum divided to shew a white deposit of growth infiltrating the subcutaneous tissue.   The surface of the deposit is warty and in parts slightly ulcerated, and histologically it has the characters of a squamous-celled epithelioma with cell-nests.

Removed by Mr. Astley Cooper from the septum of the scrotum of a chimney-sweep about the year 1805.   *See Old Museum Book*, No. 48.

2180   Epithelioma of the Scrotum.

A scrotum shewing at the upper part of the right side a large oval ulcer about two inches in its longest diameter. The edge of the ulcer is sharply defined, and its base is covered with fungating growth. By the side of the ulcer there is a large flattened nodule of growth with well-marked everted edges. The skin of the scrotum is blackened and warty. Histologically the new growth is a squamous-celled epithelioma.

Amos P., æt. 63, a chimney-sweep, was admitted under Mr. Jacobson for a swollen and ulcerated scrotum. The growth was first noticed three months before admission. There were numerous brown warts on the inner side of the thighs. The scrotum and testes were removed, and the patient died one week later. At the autopsy the liver was found to be in a condition of early cirrhosis. There were no secondary deposits. *See Insp.* 1886, No. 302.

Section XL.—INJURIES AND DISEASES OF THE URETHRA.

2181   Rupture of the Urethra.

The bladder and penis of a child mounted to shew complete transverse laceration of the membranous portion of the urethra just in front of the prostate.

Edwin S., æt. 10, was admitted under Mr. Hilton in a moribund condition, having been run over by a tramway-car. At the autopsy the pelvis was found to be fractured, and there was a lacerated wound in the perinæum. *See Insp.* 1861, No. 157.

2182   Rupture of the Urethra.

The bladder and penis of a youth, mounted to shew an almost complete transverse laceration of the urethra immediately in front of the prostate. On the reverse of the specimen the surrounding tissues are seen to be in a condition of suppurative cellulitis.

Y 2

2183   Ruptured Urethra.   Retro-vesical Abscess.

A bladder with portions of the penis and scrotum, mounted to shew a complete transverse laceration of the urethra half an inch in front of the verumontanum. The surrounding tissues are sloughing, and there is a considerable abscess cavity behind the bladder, which communicates by a large opening with the rectum three inches above the anus.   The bladder is in a condition of membranous inflammation.

William C., æt. 18, was admitted under Mr. Hilton for injuries received by being crushed between a cart-wheel and a wall. The urethra was found to be lacerated, and subsequently urine escaped by the rectum. The patient died a month after the accident, and at the autopsy the pelvis was found to be fractured and there were scattered nodules of hepatisation in the lungs. *See Insp.* 1861, No. 94.

2184   Urethra punctured per Rectum.

A bladder with portions of the urethra and rectum, mounted to shew a communication produced by a trocar between the rectum and the prostatic urethra.   The instrument has entered the urethra just to the left of the verumontanum, the canal at this point being considerably dilated.   There is a stricture of the spongy urethra, and a false passage leading from the dilated portion of the canal into the cellular tissue in front of the bladder.   Suppuration extends from this point to the connective tissue between the bladder and the rectum.

Charles J., æt. 31, was admitted under Mr. Cock in 1847 with an abscess at the neck of the bladder and a stricture of the urethra, through which a catheter could not be passed. A trocar was passed through the rectum in the direction of the bladder, and a small quantity of urine was drawn off. The patient died from suppurative pyelo-nephritis and peritonitis. *See Insp.* vol. 35, p. 41 ; and *Med.-Chir. Trans.* vol. 35, p. 184.

2185   Perforation of the Urethra. Recto-vesical
        Abscess.

The male pelvic viscera seen from the right side, partially dissected to shew between the bladder and the
rectum, below the situation of the recto-vesical pouch,
an abscess cavity of considerable size. A rod has been
passed from this cavity through a circular perforation
measuring a quarter of an inch in diameter and opening
into the membranous urethra immediately in front of
the prostate. The ureters are thickened and dilated.

Presented by Mr. Aston Key.

2186   Perforation of the Urethra. Retro-vesical
        Abscess.

A bladder with the prostatic urethra, shewing on the
floor of the latter a perforation leading into a considerable abscess cavity situated between the bladder
and rectum. The neighbouring connective tissue is in
a condition of suppurative inflammation.

John C., æt. 18. *See Insp.* 1896, No. 17.

2187   Perforations of the Urethra. Penile Abscess.

A bladder with the urethra laid open to shew a condition
of patchy membranous urethritis. On the floor of the
spongy portion of the canal and on its sides are several
openings of considerable size leading into the surrounding
tissues, which are necrotic and gangrenous. The bladder
is hypertrophied.

Richard L. was admitted under Mr. Bransby Cooper in 1826
with difficulty in micturition, for the relief of which he had been in
the habit of passing bougies. An abscess which was found pointing
at the anterior part of the scrotum was incised. Three weeks later
the patient died, and at the autopsy the abscess was found to extend
from the glans to the root of the penis. *See Insp.* vol. 1, p. 143.

2188   **False Passages in Urethra.   Puncture per Rectum.**

A bladder with portions of the urethra and rectum laid open to shew the track of a puncture which enters the bladder on its posterior wall midway between the summit and its neck, half an inch to the right of the median line. The course of the puncture is oblique, and its lower orifice is seen in the rectum two inches above the anus. There are numerous false passages in the urethra.

William S., æt. 26, was admitted under Dr. Gull with traumatic paraplegia of four days' duration, associated with retention of urine. A catheter was daily employed, and the bladder was ultimately punctured per rectum. He died one month after admission, and at the autopsy the spinal cord was found in a condition of hæmorrhagic myelitis and there was cellulitis due to extravasation of urine. *See Insp.* 1862, No. 22.

2189   **Traumatic Stricture of the Urethra.   Perinæal Section.**

A bladder with the penis and a portion of the rectum mounted to shew an impermeable annular stricture at the bulbous portion of the urethra. About an inch in front of the verumontanum is seen the opening left after Cock's operation.

From a youth, æt. 16, who fell astride and ruptured his urethra. About six weeks after the accident he began to suffer from symptoms of stricture, for the relief of which perinæal urethrotomy was performed. About ten days later the patient died from pyæmia.

Presented by Mr. Cooper Forster, 1866.

2190   **Traumatic Stricture of the Urethra.   Prostratic Abscess.**

A bladder with the urethra laid open to shew a short stricture situated at the commencement of the spongy portion of the canal, and in the recent state barely admitting a small probe. There is considerable hypertrophy of the walls of the bladder and the proximal portion of the urethra is dilated. On the reverse of the specimen the ureters are seen to be thickened and dilated, and there is an abscess cavity in the right lobe of the prostrate.

Joseph B., æt. 20, was admitted under Mr. Hilton for stricture of the urethra following upon an accident six and a half years previously. Three weeks after admission he died, and at the autopsy the kidneys were found to contain many small abscesses, and the pelves and ureters were in a condition of membranous inflammation. *See Insp.* 1862, No. 92.

2191   Stricture of the Urethra.

A portion of a bladder with the urethra laid open to shew at the bulb a tight annular stricture, immediately beyond which is seen on the floor of the canal a valvular fold of mucous membrane which was thought to have been produced by the passage of a catheter.

Richard B., æt. 45, was admitted under Dr. Golding Bird in 1850 for Bright's disease. Nine days later he died and at the autopsy the kidneys were found to be small and granular. *See Insp.* vol. 36, p. 90.

2192   Stricture of the Urethra.

A bladder with the urethra laid open to shew at the commencement of the spongy portion of the canal a tight stricture an eighth of an inch long, by the side of which is seen a false passage about half an inch in length. On the vesical side of the stricture the urethra is much dilated. The bladder is hypertrophied.

2193   Stricture of the Urethra.

A penis with the urethra laid open to shew the three inches of the canal immediately behind the corona narrowed and contracted by a deposit of fibrous tissue in its wall. Beyond the stricture the lumen of the urethra is considerably enlarged.

Timothy M., æt. 55, was admitted under Mr. Bransby Cooper in 1844, for injuries to the elbow-joint, and died about five weeks later from pyæmia. At the autopsy the bladder was found to be hypertrophied. *See Insp.* vol. 32, p. 305.

2194   Stricture of the Urethra.   Extravasation of Urine.

A bladder with the urethra laid open to shew a stricture about an inch in length, situated in the spongy portion of the canal. Two inches and a half from the bulb immediately in front of the stricture is seen a false passage leading into a gangrenous cavity by the side of the urethra. Behind the stricture the urethra is greatly dilated and the bladder is hypertrophied.

From a patient of Mr. Aston Key. He died from extravasation of urine.

2195   Stricture of the Urethra.   False Passages.

A portion of a bladder with the urethra mounted to shew the first inch and a half of the spongy portion of the canal to be narrowed and thickened. The rods indicate two false passages, one on the floor of the urethra and one in the left lateral wall. Above the stricture the urethra is dilated and the orifices of the prostatic ducts are unusually large.

John McM., æt. 45, was admitted under Mr. Aston Key in 1847 for phthisis, and died from perforation of a tuberculous ulcer of the intestine. *See Insp.* vol. 34, p. 285.

2196   Stricture of Urethra.   False Passage.

A bladder with the urethra, in the spongy portion of which, at a distance of two and a half inches from the bulb, is situated a short annular stricture, through which a rod has been passed to indicate the course of the urethra. Immediately in front of this is a false passage measuring almost two inches in length and terminating a little to the right of the canal. It has been laid open and its interior is seen to be partially coated with granular lympth. The bladder is greatly hypertrophied.

James G., æt. 39, was admitted under Mr. Hilton for perinæal abscess, consequent upon a stricture from which he had suffered for some years. The abscess was opened and the patient died from pleuro-pneumonia. *See Insp.* 1860, No. 34.

2197   Stricture of the Urethra.   False Passages.

A bladder with the urethra laid open to shew a short stricture at the commencement of the spongy portion of the canal. At the situation of the stricture the mucous membrane is greatly thickened, whilst beyond it the canal is dilated and shews numerous short sinuses and several polypoid projections. Rods have been passed into two of the sinuses which lead through the prostate to open at the neck of the bladder. The bladder shews some hypertrophy of its walls and is sacculated.

2198   Stricture of the Urethra.   Patent Urachus.

A bladder with the urethra laid open to shew a stricture situated in the penile portion of the canal, which is much ulcerated and dilated beyond the obstruction. The bladder itself is extremely contracted, and from its summit proceeds a canal, gradually narrowing towards the point where it opens through the skin at the situation of the umbilicus. On the posterior wall of the bladder is a short sinus leading towards the recto-vesical pouch, in which a piece of intestine is adherent. The right ureter is much dilated and there is evidence of pelvic cellulitis. *See Drawing,* 365.

2199   Stricture of the Urethra.   Puncture per Rectum.

A bladder and urethra laid open to shew a stricture in the bulbo-membranous portion of the canal, and at the neck of the bladder in the median line the opening left after puncture per rectum. Corresponding to the sinus is seen in the rectum a small dark spot over which the mucous membrane is cicatrised.

John D., æt. 30, was admitted under Mr. Birkett for stricture of the urethra, for the relief of which the bladder was punctured per rectum. He left the hospital relieved and ten months later he died from chronic pachy-meningitis. *See Insp.* 1859, No. 137.

2200   Stricture of the Urethra. Puncture per Rectum.

A bladder with the urethra and a portion of the rectum laid open to shew at the bulb an annular stricture an eighth of an inch in length, beyond which the urethra is much dilated. In the trigone there is a small opening, through which a blue rod has been passed to enter the rectum three inches above the anus.

Joseph S., æt. 55, was admitted under Mr. Birkett and died six days later from suppurative nephritis. One year previously the bladder had been punctured per rectum and the cannula left in position for six days. When he left the hospital about two months later the urine was passed by the urethra, none escaping per rectum except when the bladder was much distended. *See Insp.* 1857, No. 100.

2201   Stricture of the Urethra. Puncture per Rectum.

A bladder with portions of the urethra and rectum mounted to shew the track of two punctures from the rectum to the bladder. The vesical orifices are situated just behind the prostate, one in the median line and the other about one third of an inch to the left of it. In the rectum they occupy the same relative positions and are situated about two inches from the anus. The urethra is strictured at the bulbous portion and the wall of bladder is greatly hypertrophied.

Simeon H., æt. 46, was admitted under Mr. Birkett with retention of urine, for the relief of which the bladder was punctured per rectum a week before his death. Similar operations had been performed about a year previously. The patient died from suppurative nephritis. *See Insp.* 1861, No. 31.

2202 Stricture of the Urethra. Cock's Operation.

A bladder and urethra laid open to shew in the prostatic urethra, immediately in front of the verumontanum, the opening left after Cock's operation. The opening is half an inch in length, and on the reverse of the specimen is seen to correspond with a wound in the perinæum immediately in front of the anus. There is a short impermeable stricture in the penile urethra, two inches from the meatus urinarius. The bladder is greatly hypertrophied and dilated.

George P., æt. 62, was admitted under Mr. Davies-Colley for retention of urine, having been under treatment six years previously for stricture of the urethra. Perinæal urethrotomy was performed and the patient died some hours later. At the autopsy the left kidney was found to be hydronephrotic and the right was in a condition of early pyelo-nephritis. *See Insp.* 1892, No. 139.

2203 Stricture of the Urethra. Perinæal Fistulæ. Cock's Operation.

A bladder with the urethra and a portion of the perinæum partially dissected to shew in the floor of the urethra an opening about three inches in length, commencing a little in front of the prostate. On the right side of the membranous portion of the canal is a sinus, indicated by a yellow rod, which in the recent state communicated with a gangrenous cavity by the side of the bladder. On the reverse of the specimen is seen a partially cicatrised opening through the perinæum measuring two inches in depth, and presenting on its lateral walls several fistulæ leading into the surrounding fibro-fatty tissue.

William G., æt. 66, was admitted under Mr. Golding-Bird for urinary fistulæ in the perinæum, associated with long-standing stricture of the urethra. Thirty-six years previously Mr. Cock had performed the operation which bears his name. After an unsuccessful attempt to perform internal urethrotomy, Mr. Golding-Bird made an incision into the urethra through the middle line of the perinæum, and the patient died three days later. *See Insp.* 1894, No. 169.

2204   **Dilated Stricture of the Urethra.**

A bladder with the urethra laid open to shew in the bulbous portion a roughened and thickened condition of the mucous membrane, the calibre of the tube at this point being little, if at all, diminished.

William F., æt. 36, was admitted under Mr. Hilton for stricture of the urethra, which was speedily relieved. He was about to be discharged after a week's treatment when he died from peritonitis, resulting from the perforation of a typhoid ulcer in the ileum. *See Insp.* 1864, No. 219; and *Prep.* 875.

2205   **Urethral Fistula.**

A glans penis shewing close to the frænum an excavated ulcer, in the base of which is a sinus communicating with the urethra. There is a smaller and more superficial ulcer on the opposite side of the frænum.

John A., æt. 55, was admitted under Mr. Aston Key in 1843 for stricture of the urethra, from which he had suffered for thirteen years. There was an abscess in the perinæum. At the autopsy the kidneys were found to be in a condition of suppurative nephritis. *See Insp.* vol. 32, p. 265.

2206   **Peri-urethral Abscess.**

A urethra with the bladder laid open to shew, a little in front of the bulbous portion, a sinus leading into an abscess situated on the left side of the tube. The middle lobe of the prostate is much enlarged and presents several perforations produced by the passage of catheters. *See Model,* 100 (5).

Presented by Mr. Bransby Cooper.

2207   **Polypus of the Urethra.**

A bladder with the urethra laid open to shew attached to the floor of the canal at the situation of the bulb a polypus measuring half an inch in length and resembling in size and shape a melon-seed.

William R., æt. 40, was admitted under Mr. Aston Key in 1827 with symptoms suggesting stricture of the urethra, having been treated at the hospital on a previous occasion for similar symptoms. He died suddenly from the rupture of an aortic aneurysm into the left pleural cavity. *See Insp.* vol. 3, p. 17 ; and *Prep.* 1454 [2nd Edit.].

2208   Epithelioma of the Urethra.

A penis with a portion of the bladder laid open to shew a malignant growth involving the first three inches of the spongy portion of the urethra. The growth appears as an elongated ulcer having fungating nodules in its base. The corpora cavernosa are invaded and there is a prominent mass of malignant deposit between the urethra and the skin of the perinæum, in which is seen the opening made by operation. Histologically the growth consists of a fibrous stroma, the alveoli of which are filled with large squamous epithelial cells.

Thomas C., æt. 55, was admitted under Mr. Howse with a fluctuating swelling in the anterior part of the perinæum. Pressure on the swelling produced a discharge of pus from the urethra. Seven months before admission the patient first noticed difficulty in micturition, for the relief of which he was treated by the passage of catheters. Mr. Howse incised the swelling and three days later the patient died. At the autopsy the bladder was found to be in a condition of phlegmonous cystitis, and there was early suppurative peritonitis. There were secondary deposits of growth in the aortic glands. *See Insp.* 1876, No. 41.

2209   Urethral Calculus.

A spherical calculus measuring a quarter of an inch in diameter, which was extracted from the urethra. The nucleus and crust are dark brown in colour and are separated from each other by a zone of a lighter material. The exterior is brown and polished, and the calculus is said to consist of oxalate of calcium and animal matter.

Presented by Mr. Talent.

2210   Urethral Calculus.

Three small masses of phosphatic material which formed part of a urethral calculus, the nucleus of which was a straw.   *Removed by Sir Astley Cooper.*

2211   Urethral Calculus.

A cylindrical calculus measuring a third of an inch in length and an eighth of an inch in transverse diameter. It has a white colour and a somewhat irregular surface. One end is rounded, the other truncated.

It was removed by incision from the urethra just below the glans. The patient was a boy æt. 2¼ years.

2212   Urethral Calculus.

An ovoid calculus measuring half an inch in length and an eighth of an inch in diameter, which was removed from the urethra by Mr. Aston Key. Its body is of a light brown colour and consists of thin concentric laminæ of oxalate of calcium. There is a partial outer crust consisting of a white deposit of fusible phosphates. It weighs 11½ grains, and was analysed by Dr. Babington.

2213   Urethral Calculi.

Two small ovoid calculi, which were removed from the urethra by Mr. Sudlow Roots. Exteriorly they present a white roughened surface, and on analysis were found to consist of fusible phosphates.

2214   Urethral Calculus.

The half of a cylindrical calculus measuring half an inch in length, which was removed from the urethra. The cut surface shews the concretion to be of a uniform brown colour and of rather loose texture. It was analysed by Mr. Brett, who found it to consist in the main of uric acid with traces of oxide of iron and urates of sodium and calcium.

2215   Urethral Calculi.

Three cylindrical calculi, the largest of which measures an inch in length. Two of them have blunt extremities, whilst the ends of the other are pointed. Their exterior is brown with a slight coating of white phosphatic deposit.

Removed by Sir Astley Cooper from the urethra.

2216 Blood-clots from the Urethra.

Two blood-clots of cylindrical form measuring about an eighth of an inch in diameter, which, with many similar casts, were removed per urethram.

From a man, æt. 26, who for some years had suffered from albuminuria with occasional attacks of hæmaturia. At the autopsy both kidneys were enlarged, the right weighing 44 ounces. They were affected with cystic disease. *See Note-Book*, No. 2, p. 26.

Presented by Dr. Iliff, 1837.

2217 Foreign Body from the Urethra.

A piece of French chalk an inch in length and a quarter of an inch thick attached by a piece of string to a shoe-horn.

Charles L., æt. 18, was admitted under Mr. Callaway in 1856. He had inserted the piece of chalk into his urethra with a string attached, and being unable to remove it, tied the shoe-horn to the other end of the string. The foreign body was removed through an incision in the median line.

2218 Foreign Body from the Urethra.

A pin an inch and a quarter long, which is encrusted, except at its point, with phosphates.

George B., æt. 10, was admitted under Mr. Bransby Cooper in 1850 with symptoms resembling those of stone in the bladder. The pin was found and removed from the membranous portion of the urethra.

2219 Foreign Body from the Urethra.

The wooden handle of a pen-holder about five inches long and an eighth of an inch in diameter, which was removed from the male urethra by Mr. Birkett.

2220   Broken Catheter removed from the Bladder.

The two halves of a broken flexible metal catheter, the lower of which, seven inches in length, was removed from the bladder.   *Presented by Mr. Aston Key*, 1825.

2221   Pieces of a Catheter extracted from the Bladder.

A gum elastic catheter with its stilet. At the lower end of the instrument are three fragments, each of them about an inch in length, the two upper ones being partially coated with white phosphatic deposit.

Alfred B., æt. 24, was admitted under Mr. Hilton in June 1858, having broken off in his bladder the extremity of a catheter which he was in the habit of passing " for the relief of an imaginary stricture." The broken portion was removed in pieces on three successive occasions by means of a lithotrite. The patient made a good recovery.

2222   Catheter mended with Thread.

A flexible metal catheter mended in three places with pack thread, in which state it had for some time been used by a tailor.

Section XLI.—INJURIES AND DISEASES OF THE PENIS.

Amputation: 2223, 2224.
Phimosis: 2225.
Œdema: 2226.
Gangrene: 2227-2232.
Chancre: 2233-2235.
Warts: 2236-2238.
Epithelioma: 2239-2246.
Melanotic Growth: 2247.
Sebaceous Cyst: 2248.
Preputial Calculus: 2249.

2223   Self-inflicted Amputation of External Genitals.

A penis and testes, on which are seen small black fragments of coal and cinder.

> From Robert B., who, " after attending a Bethel Union on board a collier at Erith, in a fit of religious melancholy, amputated his genital organs with a razor and threw them under the galley fire."

2224   Amputation of the Penis.

The cicatrised stump of a penis shewing the orifice of the urethra, preserving its natural relation to the corpora cavernosa.

> John B., æt. 50, was admitted under Mr. Hilton with epithelioma of the penis. Amputation was performed in such a way that the corpus spongiosum was half an inch longer than the corpora cavernosa. The patient subsequently died with secondary deposits and cancerous ulcers in the groin. *See Insp.* 1856. No. 59 ; and *Drawings*, 447 (50 & 51).

2225   **Phimosis.**

A prepuce, the orifice of which is narrowed so as to admit only a fine probe. The mucous surface is adherent to the glans penis.

2226   **Œdema of the Prepuce.**

A prepuce enlarged by œdema so as to measure three quarters of an inch in thickness.

Presented by Mr. Morgan.

2227   **Sloughing of the Prepuce.**

A penis mounted to shew the condition resulting from sloughing of the prepuce and the adjacent integuments. The corpora cavernosa are shrunken and denuded for a distance of two inches from the corona.

2228   **Gangrene of Penis and Scrotum.**

The entire integuments of the penis and scrotum of a child separated as a sphacelus.

William II., æt. 10 months, was admitted under Mr. Davies-Colley in 1884 with gangrene of the scrotum and penis. He was an illegitimate child and had suffered much privation. The inflammation of the genitalia was first noticed by the mother five days before admission. The slough which forms the preparation separated fourteen days from the commencement of the illness. The child was discharged in good health four months after admission. *See Surgical Reps.* vol. 115, Case 212.

2229   **Gangrene of the Penis.**

A penis, partially injected, mounted to shew the skin and subcutaneous tissues greatly swollen and in a condition of diffuse gangrenous cellulitis. The glans and the corpora cavernosa appear to be unaffected.

2230   **Gangrene of the Penis.**

The pubes with the scrotum and a portion of the bladder of an adult male. The penis is absent and the root of the organ is represented by a smooth cicatrised surface, at the bottom of which is seen the slit-like opening of the urethra. The halves of the scrotum occupied by the testis form prominent folds on either side somewhat resembling the labia majora of the female.

William J., æt. 43, was admitted under Dr. Addison with phthisis and lardaceous disease, from which he died. Twenty years previously he had been in the hospital under Mr. Aston Key for phagedæna resulting in the complete destruction of the penis. *See Insp.* 1857, No. 133.

2231   Sloughing of the Penis.

The dorsal half of a glans penis, including the meatus urinarius, which separated as a slough. The surface, which is of a brown colour, is eroded by ulceration.

From a patient, æt. 79, who was admitted under Mr. Birkett in 1854, suffering from gangrene of the penis, following upon a blow inflicted six weeks previously. After the separation of the sphacelus the patient made a good recovery.

2232   Cellulitis of the Corpus Spongiosum.

Portions of a penis and bladder laid open to shew the corpus spongiosum for a distance of three inches from the bulb in a condition of gangrenous cellulitis. The mucous membrane of the urethra is almost completely destroyed, and just in front of the prostate is seen an opening in the floor of the canal leading into a sloughy cavity in the muscular tissues.

Henry F., æt. 52, was admitted under Mr. Birkett with cellulitis of the scrotum, the symptoms resembling those of extravasation of urine. A catheter was easily passed. The scrotum was incised, and three days later the patient died. At the autopsy the genitalia and the cellular tissue at the lower part of the abdomen were gangrenous. *See Insp.* 1873, No. 182 A.

2233   Chancre of the Penis.

A portion of the skin of a penis shewing an oval chancre an inch in length and projecting a quarter of

an inch above the surrounding tissues. The summit of
the chancre is flattened and presents a granular surface
from partial removal of the epithelium. A section has
been taken from the edge of the sore and shews the
corium to be densely infiltrated with small round cells.

From a patient who had noticed the chancre for two weeks.
It was excised by Mr. Tweedie in 1840.

2234   **Chancre of the Penis.**

A prepuce, the extremity of which is greatly enlarged
and œdematous. On the dorsum is seen a small excavated ulcer with somewhat thickened edges.

* From a patient who suffered from phimosis and whose prepuce
was amputated for the cure of a venereal sore by Mr. Morgan.

2235   **Venereal Sore on the Prepuce.**

A portion of the skin of a penis with the prepuce
mounted to shew on the under surface of the latter a
shallow ulcer of irregular shape measuring an inch in
its longest diameter.

From a patient who suffered from phimosis and whose prepuce
was removed by Mr. Aston Key.

2236   **Warty Growth of the Prepuce.**

A prepuce, the free border of which is the seat of a
considerable number of closely aggregated cauliflower-
like excrescences having the histological characters of
papillomata.   *Presented by Mr. Aston Key.*

2237   **Warty Growth of the Prepuce.**

A prepuce shewing a collar of growth springing from
its free margin and consisting of closely aggregated
slender processes. Histologically the structure is that
of a simple papilloma.

2238   Papilloma of the Penis.

The extremity of a penis, the prepuce of which is enormously thickened and covered by a cauliflower-like growth with overhanging edges. The growth appears to have sprung from the mucous surface of the prepuce and forms a complete broad collar around the glans, the surface of which is itself the seat of similar deposit. Histologically the structure is that of a papilloma with no evidence of malignancy. The submucous tissue is densely infiltrated with small inflammatory cells.

2239   Epithelioma of the Penis.

The extremity of a penis divided longitudinally to shew the glans and corpora cavernosa infiltrated by a soft white growth which projects as a prominent mass at the situation of the fraenum. The prepuce, which completely covers the glans, is partially adherent to the growth and infiltrated by it.

Presented by Mr. Jacobson, 1895.

2240   Epithelioma of the Penis.

The extremity of a penis mounted to shew the glans enlarged and distorted by an irregular mass of malignant deposit, the surface of which is in parts warty and in parts excavated by ulceration. The growth, which extends to the skin of the prepuce and invades the corpora cavernosa, has the histological characteristics of a squamous-celled epithelioma.

The penis was amputated by Mr. Astley Cooper about the year 1807.

2241   Epithelioma of the Penis.

The end of a penis shewing the glans partially destroyed by epitheliomatous ulceration. The surface is covered with warty growths which extend on the dorsum of the organ to the corona and the adjacent edge of the prepuce.

2242   Epithelioma of the Penis.

Portions of a penis and scrotum mounted to shew large masses of cauliflower-like growth having the histological characters of a squamous-celled epithelioma with numerous cell-nests.

From a patient of Mr. J. Morgan. The penis was amputated for a malignant growth which reappeared in the stump.

2243   Epithelioma of the Penis.

The end of a penis divided longitudinally to shew the fraenum and the adjacent parts of the prepuce infiltrated by a firm white growth, the surface of which is covered with short warty excrescences. The deposit extends into the glans nearly as far as the fossa navicularis. Histologically the growth is a squamous-celled epithelioma without cell-nest formations.

Presented by Sir Astley Cooper.

2244   Epithelioma of the Penis.

The extremity of a penis partially injected and mounted to shew the glans and a portion of the prepuce covered by a cauliflower-like mass of growth having the histological characters of a squamous-celled epithelioma.

2245   Epithelioma of the Penis.

The extremity of a penis mounted to shew a malignant growth which has fungated through the prepuce and appears as a mushroom-shaped mass surrounded by skin. On the reverse of the specimen it is seen to be continuous with a similar warty growth which lines the mucous surface of the prepuce and extends for a short distance

upon the corona. Histologically the deposit has the
character of a squamous-celled epithelioma with cell-
nests.

From a patient, æt. 65, who was admitted under Mr. Birkett
in 1854. The disease was first noticed on the inner surface of the
prepuce.

2246   **Epithelioma of the Penis.**

A glans penis with the prepuce laid open to shew a
warty growth which entirely covers the glans and those
parts of the foreskin with which it is in contact. Histologically the growth is a squamous-celled epithelioma
with many cell-nests.

Henry W., æt. 28, was admitted under Mr. Hilton in 1856.
He had suffered for twenty years from partial phimosis resulting
from injury. Six months before admission he noticed a wart
between the prepuce and the glans. The penis was amputated
and the patient made a good recovery.

2247   **Melanotic Growth of the Glans Penis.**

A portion of a glans penis mounted to shew on one side
of it a black deposit infiltrating the organ and slightly
projecting above the surrounding surface. There are
several pigmented patches in its neighbourhood, some
of which are superficial, whilst others are seen on the
cut surface to extend more deeply. Histologically the
growth consists of large spheroidal cells loaded with
pigment and arranged in a scanty alveolar stroma.

Robert R., æt. 43, was admitted under Mr. Golding-Bird in
1881 for a melanotic growth of the glans penis which had been
first noticed twelve months previously. The portion of the glans
affected was removed and the patient was in good health fourteen
months after the operation. *See Surgical Reps.* vol. 93, Case 26.

2248   **Sebaceous Cyst of the Prepuce.**

The prepuce of a child shewing on its under surface
close to the median raphe a small nodule a quarter of

an inch in diameter which has been incised to shew a smooth-walled cavity from which sebaceous material has been removed.

From a boy, æt. 9. The nodule had been noticed for five years. He was circumcised by Mr. Bryant in 1862.

2249   **Preputial Calculus.**

The dried extremity of a penis mounted to shew a concretion lodged within the prepuce. The stone is white in colour, ovoid in shape, has a roughened surface, and measures half an inch in its longest diameter.

PRINTED BY TAYLOR AND FRANCIS, RED LION COURT, FLEET STREET.